CW00691215

FEMINISM, ANIMALS AND SCIENCE

The Naming of the Shrew

Lynda Birke

OPEN UNIVERSITY PRESS
Buckingham • Philadelphia

Open University Press
Celtic Court
22 Ballmoor
Buckingham
MK18 1XW

and
1900 Frost Road, Suite 101
Bristol, PA 19007, USA

First Published 1994

Copyright © Lynda Birke 1994

All rights reserved. Except for the quotation of short passages for the
purpose of criticism and review, no part of this publication may be
reproduced, stored in a retrieval system, or transmitted, in any form
or by any means, electronic, mechanical, photocopying, recording or
otherwise, without the prior written permission of the publisher or a
licence from the Copyright Licensing Agency Limited. Details of such
licences (for reprographic reproduction) may be obtained from the
Copyright Licensing Agency Ltd of 90 Tottenham Court Road, London,
W1P 9HE.

A catalogue record of this book is available from the British Library

ISBN 0 335 19197 5 (pb)   0 335 19198 3 (hb)

Library of Congress Cataloging-in-Publication Data
Birke, Lynda I. A.
    Feminism, animals, science : the naming of the shrew / Lynda
Birke.
        p.   cm.
    Includes bibliographical references and index.
    ISBN 0–335–19198–3 (hardback)   ISBN 0–335–19197–5 (pbk.)
    1. Feminist theory.   2. Feminism.   3. Human-animal relationships.
4. Animal rights.   5. Ecofeminism.   6. Biology—Philosophy.
7. Science—Philosophy.   I. Title.
HQ1190.B57 1994
305.42'01—dc20   94–19790
                                                            CIP

Typeset by Graphicraft Typesetters Ltd., Hong Kong
Printed in Great Britain by Biddles Limited, Guildford and Kings Lynn

For my parents,
and
in memory of
all those wonderful creatures
whom I have had the privilege of knowing

CONTENTS

Acknowledgements

There are many sentient creatures to whom I am indebted in developing my thoughts and feelings about feminism and animals, feminism and science. Many of them are human; many, many more are not. To those who are, I can perhaps express my gratitude. But it is harder to do so for those whose first language is not human. Their alleged inability to speak is surpassed by my (our) supreme inability to listen.

Many years of thinking about feminism and biology, and discussing the issues, form the backdrop to this book. I can, of course, name many individuals who have contributed directly to that process. But the process is wider than that – it has been helped by participation in women's liberation conferences, women's self-help health groups, lesbian/feminist consciousness-raising groups; students and colleagues, especially in women's studies; environmental groups; those scientists who do try to challenge the hegemony of reductionist thinking and who share my belief that animals are not mere automata. They have all helped to weave the interconnections.

More specifically, I want to thank a number of individuals. Before I began writing, many of the ideas that form the basis for this book were discussed with Gail Vines and with Ted Benton in Britain, and with Barbara Noske in the Netherlands. My thinking also draws on research discussions focusing on the controversy surrounding animal experimentation, notably with Mary Ann Elston, Mike Michael and Jane Smith.

I am particularly grateful to several friends who have taken the trouble to read/comment on all or part of the manuscript: thanks to Ruth Hubbard, Anne Fausto-Sterling, Mike Michael, Dawn Sadler and Margrit Shildrick for spending that time.

Whatever its conception, the time in which this book was written began at a time of great pain and trauma in my life. It is thanks to the support of close friends (human and non-human) that I kept going. Not only did my life seem to be falling apart, but my old and wonderful dog, Susie, developed lymphosarcoma in the middle of it all. Her subsequent death felt too great a burden for me to bear. But it would have done her memory a disservice if I had given up thinking about feminism and animals. Special friends helped to keep me going; they kept me believing in myself and what I was doing. To them, I owe a very special thank you. At home, that emotional support through the darkest days came especially from Sandy Best, Dot Tester, Roz Heredia, Lean Heng Chan, Mike Michael, Jean Barr, Nic Fryer, and my parents. Frequent trips to the USA helped to keep me sane, especially at the point of crisis, and I want to thank Phyllis Robinson and Kathy Marmor (for reminding me that laughter heals), Ruth Hubbard, Anne Fausto-Sterling, Sandra Harding and Barbara Gates for keeping my spirits up. The migrating snow geese helped, too, Barbara.

Annie Barker reminded me of the triple meaning of 'shrew' – not only the animal, but also as a term of abuse for women in the past, and hence the title of an early magazine produced in the women's liberation movement. So the subtitle was born.

Finally, I have been privileged to share moments in my life with many non-humans. There are, of course, moments of awe and wonder when we cross paths with wild creatures. Who was watching whom, I thought, as I photographed a ground squirrel in Canada watching me? Or the secretive badgers that live close by my house? Or the snow geese migrating over the salt marshes in Delaware? And then there are the many animals that I have worked with in science; whatever the stereotypes, rats are wonderful animals and have taught me much. Many more moments have been shared with those companion animals with whom I live, dogs, horses, cats and Junior the pet chicken. They teach me about love, and about listening; they teach me about sharing. And above all, they teach me about their individuality; the relationships I have with each of them is different – and cannot be described by the limited vocabularies of science.

Part I

ANIMAL MEANINGS

1

INTRODUCTION

Prologue

I dreaded looking into his eyes . . . If I had been born into slavery,
and my partner had been sold or killed, my eyes would have looked
like that . . . Blue was like a crazed person. Blue *was*, to me, a crazed
person . . . he looked at me [with] a look so piercing, so full of grief,
a look so *human*, I almost laughed (I felt too sad to cry) to think
there are people who do not know that animals suffer.

(Walker 1988: 5)

Though I looked for sameness . . . I soon realized that I was actually
more interested in the individuality of the animals I watched. In
fact, I began to comprehend the magnitude of the role of idiosyn-
cratic behavior on the scheme of things. Diversity is actually the
creative element in the evolutionary process; whereas, the conform-
ity we like to measure is merely the 'setting' of nature's wild impulses after these have proved successful over long ages of natural
selection.

(Ryden 1972)

Horses, like Blue in Alice Walker's story, or like the wild mustangs Hope
Ryden has worked to protect, have always been my friends. My passion
for them grew alongside my passion for natural history and science. It
seems fitting, then, to begin a book about feminism, science and animals
with thoughts about horses.

In Walker's short story, she recounts her feelings about the horse
grazing in a nearby field. The horse, Blue, seemed happy when he had
a mate; and poignantly miserable when the mate was taken away. Later,

when friends visit, she sits down to dinner: 'As we talked of freedom and justice one day for all, we sat down to steaks. I am eating misery, I thought, as I took the first bite. And spit it out' (Walker: 1988).

Walker echoes here several strands that run through feminist thinking and practice. One of these is vegetarianism. For many people concerned with animal-related issues, the slogan 'Love Your Friends – Don't Eat Them' (emblazoned on a British Vegetarian Society T-shirt) is taken literally. Feminists, too have often perceived a connection between vegetarianism and women's politics (Adams 1990).

The story brings out another theme that has resonance within feminist thinking: what Walker touches on here is her sense of common suffering with the horse. It is not the (culturally accepted) sense of humans as different from other animals that moves her, but her awareness that Blue is suffering just as she would do, and just as her people did in slavery. It is similarity, not difference, that matters here.

Hope Ryden recounts, with pain and passion, the various ways in which the American mustang – that symbol of the western prairies – is besieged by farming and government interests. She has campaigned to save the horses and in doing so has studied them. Her observation makes a related point: science tends to observe animals as exemplars in their sameness. They represent 'species horse'; but what is often missing from scientific accounts of the behaviour of animals is an understanding of each one's uniqueness, of difference.

Tensions between similarity and difference are central to western feminist thinking. In the early years of the current women's movement, the 'difference' of which we spoke was primarily that we perceived between women and men; it was 'sex roles' that we railed against, and their apparent institutionalization. By contrast, we perceived common ground among women, expressed in ideas of *the* women's movement, or *the* experience of women.

That simplistic account was soon challenged, however. Challenges to the notion of a unitary women's experience came from working-class women, from black women, from lesbians; even if there is some common ground (and that remains a contested point in feminist writing) then what is now the centre of concern for feminists is difference. It is no longer only (or even) our difference from men that is an issue; it is also our different experiences from each other.

The sense of difference evoked by Ryden and Walker – humans and other animals – informs feminist thinking relatively rarely (at least overtly). Yet it is there, part of the legacy of assumptions that western feminism has culturally inherited. Broadly, there are two ways in which a human/animal distinction is evident in feminist writing. One is in the context of environmental concerns, where issues around conservation, to do with preservation of wilderness, and to do with animal rights, may be aired (see, for example, Caldecott and Leland 1982; Diamond and Orenstein 1990). These concerns are more to do with similarity than difference; we have similar needs to other species, for example, in the protection of our earth.

The second strand focuses on difference, but is rarely made explicit in feminist writing. Western feminisms are the heirs of the rationalist traditions that, among other things, have sought to *separate* humanity (indeed, it might be more accurate to use 'mankind' here) from nature. The heritage of Judaeo-Christianity, for example, has encouraged our culture to think in terms of the domination of nature, thus emphasizing our difference, and separation, from other species; it has encouraged us to think of non-humans as inevitably inferior. Science, too, has historically become associated with a view of nature as there to be controlled, to be taken apart in order to be understood (Merchant 1982). The fundamental question that asks where humans are in the order of nature is increasingly answered in terms of control and exploitation.

Feminism has tended to take on board this separation. We have, for good reasons, challenged prevalent ideas based on biological determinism, and sought to emphasize instead the ways in which our experience of gender is socially constructed. This response has been shared in the West with other social movements, particularly socialist ones, which have tended to want to deny human evolution in the zeal to emphasize the importance of cultural history (see Benton 1993). Academic social science, too, tends to dissociate itself from thinking about 'biological' history – not least because that is deemed to be the preserve of biologists.

In that denial of biology, however, we have thrown the baby out with the bathwater. It is not, Benton argues, reasonable to deny

> the main claims of the evolutionary perspective in relation to human ancestry in some primate stock, and our kinship with contemporary primates. Unless social scientists wish to stand with the flat earthers, the Inquisitors and the bible-belt creationists they have no choice but to engage with the questions posed by our animal origins and nature.
>
> (1993: 43)

So, too, must feminists, however uncomfortable that may seem to some. Gender is undoubtedly socially constructed (and intersects with other social constructions of our experience, such as those mediated through class or race), but leaving 'biology' out is problematic. For not only do our physical bodies seem to be outside this focus, but it separates us from the rest of nature – which remains 'biological' by definition.

Why animals?

In this book, I want to focus on how society constructs ideas about 'the animal', particularly in relation to feminist critiques of science. There are obviously links between such concerns and global environmental issues; how western culture views other kinds of animals is deeply tied to how it views 'nature' more generally (see Shiva 1989). In that sense, my focus on animals can be seen as a way of indirectly exploring some of these wider issues.

There is, too, a growing concern about the ways in which non-humans are treated, and increasing advocacy for the 'rights' of at least some kinds of animals. One area where this concern is most vociferously expressed is science. Some of the people involved with animal rights issues are antagonistic to science (or perhaps to the instrumentalism of western culture that science seems at times to embody, as Jasper and Nelkin, 1992, have suggested). For them, any use of animals by science cannot be condoned. Others may make a distinction between using animals for avowedly medical research (on the grounds that suffering can be justified if there is a potential benefit – to humans, largely – that might result) and product testing. The strength of feeling (and the power of the campaigns against testing) is indicated by the consumer boycotts of cosmetic products that are not labelled as 'cruelty free'.

I should stress at this point that this is not primarily a book about animals or animal rights as such; rather, my concern is to think about how we use the *idea* of 'animal' and to link that to work on feminism and science. Why, for example, has our culture been so concerned to separate ourselves conceptually (and morally) from 'other animals'? What does that mean to us? And what *is* this 'other animal' which we are so keen to deny? Why, too, has science been so concerned to maintain that separation, consistently denying for years that other species (of mammals at least) have minds or consciousness? What is the meaning of this kind of animal, both to science and to the wider culture?

There are other reasons, though, why I want to focus on how we see, or use ideas about, animals and their similarity or difference from us, and the relationship between that set of ideas and debates about feminism and science. The first is personal. Like other feminists who are also scientists, my own experience has been of unease: how can I think as a feminist while at the same time be a scientist? In the early days of women's liberation, trying to be both sometimes felt impossible; being a scientist was beyond the pale of feminist authenticity to some of the more radical voices around. Evelyn Fox Keller summed up the con- tradictions clearly when she wrote that, 'As both a feminist and a sci- entist, I am more familiar than I might wish with the nervousness and defensiveness that such a potential conflict evokes' (Keller 1992: 175).

Yet that has not been the only source of potential conflict for me. The choice to study biology was never straightforward. On one hand, I liked science and was passionate about natural history, but on the other, I recognized that doing biology would mean doing experiments with animals. I did do biology, and so had to try to live with the moral conflicts ever since. I have always had qualms about the moral 'rightness' of whatever it was I was doing, and I continue to have doubts about the ways in which science uses animals. Those may be doubts shared with many scientific colleagues for many reasons, but for me they are partly feminist doubts.

I found it peculiar that the love of nature that, in part, drove me to want to study biology seemed to be at odds with the scientific methods in which I was trained. Loving nature meant a respect for its complexity, yet to do science means to accept its reductionism. Books from my undergraduate days testify to the ways in which the living animal becomes coded as an assemblage of parts, as machine-like: 'Living Control Systems', 'Nerve, Muscle and Synapse', 'The Ovary', among others. Doing science often meant awe at the wonderful ways that such 'systems' worked; nature was, indeed, very clever. But it also seemed to mean denying the awe at the marvellous creatures that exist in the world, in all their complexity and individuality.

A second reason for my focus is that science has sought to name both animals and women. Naming, as feminists have often stressed, is a powerful process. In scientific accounts, women have too often been described as being limited by their biology; if we are named as being victims of hormones, say, by the authority of science, then it is unsurprising if some people come to believe it (and act upon it). Animals, too, are named and defined by science; there are still squabbles among taxonomists over how to name this or that creature. Science, then, would name the shrew by a Latin name, *Sorex*. But animals are also named within the wider society, given 'common' names. The shrew has thus been labelled as evil, capable of making horses lame, a doer of bad deeds, a spiteful, mean-tempered thing. As Shakespeare well recognized, these qualities were also attributed to women – 'shrewish' was undoubtedly a gendered term and one commemorated in feminism by the women's liberation magazine of the 1970s, *Shrew*. 'Shrewish' women answer back.

Scientific naming of animals gives them a species and describes them as such; who or what they are as individuals matters little for these purposes. Indeed, another source of questions in my mind comes from the obscuring of individuality to which Hope Ryden refers. As a scientist, I have studied the behaviour of animals as part of groups. This one is assumed to represent the species 'rat', that one, the species 'Bengalese finch'. Meanwhile, I was training horses; here it is not the representative

animal that I work with, but the individual, in all its glorious idiosyn-
crasy. And, like Vicki Hearne (1987), I suspect I know more about the
species horse – *Equus caballus* – than I do, in practice, about either finches
or rats.

A central plank of feminist criticism of the sciences has been that
science has, generally, had rather little respect for people. We have
focused, for example, on derogatory statements made by biologists about
women's nature; we have joined forces with critics of medicine and the
way that poor black women have sometimes been coerced into taking
contraceptive drugs; we have questioned the ways in which women's
reproductive choices are increasingly constrained by high-tech medicine.
But science also has little respect for the non-humans it studies.

At home, I may feel that I have respect for my dogs as individuals, or
individually for the horses that I train to run and jump. In the laboratory,
however, animals tend to become numbers and codes; rat 23/5A/F is
just another rat in the experimental design. Yet somehow working with
these animals in the laboratory in studies of their behaviour, I could
never escape the feeling of wondering just what it was that made that
particular rat tick. What *was* her experience of life with her friends?
How was that experience structured by the small laboratory cages, by
the rigidly imposed cycle of artificial light and dark?

To many people – such as those who defend animal rights, or those
who train animals – the subjectivity of at least some animals is self-
evident. To many others, it is equally evident that only humans can
have subjectivity. The question is important because the debate about
the ethics of using animals in ways that might cause suffering – in
scientific experiments, for example – hinges on how much they can in
fact *experience* suffering. To experience suffering fully, some philosophers
have argued, requires subjectivity (see Leahy 1991). The question is also
important precisely because the way that science is designed expressly
denies such traits as individual subjectivity.

It is also important for us to think about our relationship to non-
human animals in relation to changing beliefs about them. Science has,
in practice, been founded on *using* animals, largely as a means to an end
(even if, ironically, many scientists entered biology in part because of
their love for natural history). It has not generally had much respect for
other creatures; we can only speculate what science might have been
like had it grown up in another culture, one in which respect for the
creatures or plants studied was assumed. Attitudes are nonetheless
changing. In Britain, for instance, the law governing the ways in which
scientists can use animals was changed in 1986, partly in response to
changing public opinion. And more scientists now voice their beliefs that
at least some kinds of animals are more sentient, or might have a 'point
of view' (Bekoff and Jamieson 1991).

The biological sciences are built on knowledge derived from animals; for me, that is central to my thinking about feminist criticism of science. Just as I have tried to grapple with the contradictions of doing science as a feminist, so too did I struggle with the contradictions of using animals in the lab (albeit in non-invasive behavioural studies). While I did so, I went on loving my dogs and horses, being a vegetarian, and trying to avoid at least some commercial products that might have been tested on animals.

Although my focus is science, I do not think that these difficult contradictions are unique to science. Vegetarianism and boycotts of particular products are both fairly common practice in feminist communities in Europe and the USA, yet how many of us avoid all drugs, or never use photographic film (which uses animal-derived gelatine)? It is, I think, important to recognize the contradictions, even while we continue to live with them. In my life, those contradictions centred on reading and participating in a science that so clearly has little respect for the integrity of the individuals, of whatever species, it studies.

Feminism, science and animals

Why is the question of human relationship(s) to animals relevant to feminist thinking about science? There are several ways in which that question could be addressed. The first is simply that animals, despite being part of the nature that science claims to study, are largely missing from feminist debate. Feminist analyses of science have not surprisingly focused primarily on issues to do with women and/or gender. One broad strand, for example, addresses equity issues: why are there relatively few women making science? What is it about science – or women's responses to it – that make it unattractive to women (or to other under-represented groups)? And why so few in the history of science and technology? Or why so few that we have heard about (see Alic 1984; Schiebinger 1989)? The central theme here is who does science; a concern with equity issues may (or may not) be linked in feminist critiques with a concern with what science itself does (see Rosser 1989).

A second theme is more concerned with what science does, and how it has come to be constructed. Scientists, within this broad view, do not simply observe nature; they also bring to their task a complex set of ideologies and assumptions derived from the wider society. Feminist writing has thus looked, for example, at the various ways in which science has dealt with women or women's bodies, and at how this literature incorporates cultural beliefs about women (Bleier 1984; Fausto-Sterling 1985; Birke 1986; Hubbard 1990). Another strand within this broad critique has focused on the epistemology of science, exploring for example whether women can have a particular standpoint on knowing, or more particularly on knowing science. One theme that has emerged in this vein is exploration of the idea that there might, or might not, be a possible feminist successor science to replace or develop from the much-criticized science that we have now (Rose 1983; Harding 1986, 1991).

Yet despite the fact that much of the feminist literature on science focuses on biology, and despite the fact that much biological knowledge inevitably depends on information gained from animals, animals are scarce in that writing. The relationship between science and the animals it often studies is rarely sufficiently problematized. The exceptions are those writings that focus specifically on some aspect of the study of animal behaviour (ethology or sociobiology, for example, or specific species such as primates; Haraway 1989b), or occasional references to the use of animals in scientific research (for example, Halpin 1989).

One reason why our relationship to animals matters to feminist theorizing has to do with how science uses animals, both literally and metaphorically. Animals are, after all, used in science in great numbers. They are used to test drugs, cosmetics and other products, to pioneer new surgical and other medical techniques in basic research. They may be studied in their own right, as exemplars of a particular species occupying a particular ecological niche. Or, more commonly, they may be used instead of humans, as 'animal models' of what scientists believe are more general physiological principles.

We might, of course, ask whether this matters to feminist critiques – or even whether it is politically acceptable to make the move (as I have done above) from talking about groups of people to laboratory animals. As I have noted, feminist work on science has tended not to address many of the issues raised by (say) the animal rights or environmental movements. One reason why feminists have not been quick to debate the ethics of animal use in science has to do with our unease about talking about animal issues at all, for it has too often been the case that those people lacking power have been derogated by likening them to 'animals'. To be likened to 'an animal' in our culture is to be diminished, or to be mindlessly out of control, and who wants to be like that?

All human cultures use various metaphors of animals and animal society. Western culture, and the science that is part of it, is no exception. How we view animal societies inevitably reflects the ways in which we experience our own: thus, as Donna Haraway has pointed out, primatology is politics by other means (Haraway 1989b). The study of primate societies has incorporated the social and political assumptions of gender, race and class that inform the wider culture, and scientific accounts of other species which build upon cultural beliefs about gender and sexual orientation. Scientific accounts in turn feed back into the wider cultural consciousness. Television natural history programmes, for example – increasingly popular viewing – not only tell us about the quaint goings-on of exotic or endangered animals, they also incorporate and reinforce prevailing beliefs about masculinity and femininity (Crowther 1994). Is it 'really' the giant panda or the baboon troupe that we are witnessing in these dramas – or yet another set of visual representations of gender or sexuality akin to those we see every day in advertising and magazines?

Another area in which we might make more connections between our theorizing about science and other politics and social movements concerns environmental issues, and there has been a surge of interest in ecofeminism in recent years. How western culture constructs its relationship to nature is, as many critics have argued, deeply problematic. It focuses on notions of dominance over and exploitation of nature that are causing widespread environmental havoc (Shiva 1989; Cox 1993). 'Nature' in this context of dominance is partly non-human nature – the other species of animals, the plants, the microorganisms, the geological structures of our earth – but it is also significantly much of humanity. Western imperialism and global exploitation assumes that it can appropriate nature's resources without significant consequence; those resources may *be* other peoples, or they may be directly affected by western destruction of their local environment. It is in that vein that, for example, the Women's Environmental Network in Britain has investigated the chocolate industry, which not only destroys the environment but also has repercussions on human lives.

A question that I particularly want to explore, in the last part of this book, begins with feminist critiques of biological determinism. Like many others, I have spent years (and many hours of wordprocessing) criticizing naive biological accounts of women's (or anybody else's) behaviour and capabilities. Biological accounts typically see gender as fixed – and fixed within cultural stereotypes. Feminists have tended, not surprisingly, to object to both the stereotypes and to the portrayal of them as biologically inherent. To the biological model, we have opposed arguments based on the social construction of gender. But that opposition, important though it has been for the development of feminist thinking about gender, is founded upon unquestioned assumptions about what we mean when we talk about the 'biological'. This in turn depends upon making assumptions about animals/nature that, I will argue, are problematic for feminism. For what we have implicitly accepted in feminist work is the notion that animals are little more than their biology; this is what constitutes their animalness. But if so, then we are inevitably going to face problems; analogies *are* drawn between human society and that of other species (and feminists sometimes do that too). So, if animals are 'mere' biology, puppets of their genes, then there will inevitably be inferences made about the mere biology at the heart of human nature.

Whatever the merits of critiques of biological determinism, they do have implications for our views of animals. Implicit in these critiques is the notion that *animal* behaviour is largely innate, to some extent determined by their intrinsic biology, while ours seems somehow to be completely emancipated from biology. Now I do not intend here to collapse human behaviour back onto determinism; the complexities of human culture argue against anything so simplistic. But I would argue for some kind of middle ground: it would be incredible if our own species' biological history was completely irrelevant (unless we are to assume the mantle of god-like disembodied minds). Equally, there is good evidence that at least some behaviour of some other species is not as fixed by biological dictates as the dichotomy between biological determinism and social construction sometimes implies.

Entrenched in the biological/social and animal/human dichotomies lies a belief in animals as 'other' (Halpin 1989). This is not only to say that we humans are not like animals; it is also to project onto them an alter ego (Burt 1988), to see them as alien. It is also to see them as part of the nature that humans must transcend. This sense of animals as 'other' permeates the history of western thought, which has, since the Enlightenment, emphasized appropriation and mastery of (alien) nature as central to human ascendancy.

Beliefs in animals as 'others' have an ideological role, just as the notion of woman as other has done. On the positive side, separating ourselves from the rest of animalkind is usually intended to give humans dignity, to expunge the 'bestial' from our natures (see Midgley 1978; Clark 1982). It is undoubtedly a way of challenging biological determinism, by emphasizing the significance of human history and culture, and the active role of people in shaping both; animals as others are outside the sphere of culture. There is much politically to be gained by this stress on cultural history, not least because it allows the possibility of active change in our social future.

Isn't there a problem here for feminists, however? While we are busily trying to claim our place in cultural history, we implicitly deny non-humans theirs. 'Othering' is tricky, however, for haven't women been seen as 'others' as a way for men to separate themselves off from a devalued and derided femininity? Is human dignity versus the animal world so different in the way that 'others' become represented?

Animals, in this formulation, tend to remain outside the realms of moral concern, becoming the repository of passive unreason. Discontinuity, moreover, rather leaves unanswered the question of what we do with human bodies and their processes, while portraying animals as merely inferior, little more than machines. Indeed, they seem at times to *be* their bodies. In moral terms, this tendency leaves animals out in the cold, not morally significant because they are held either not to reason at all or to have so little reason that they do not count for inclusion in moral arguments. Unreasoning, mere bodies, lacking language – non-humans cannot be taken seriously, some would say.

This, partly, is the basis of what has been called 'species loyalty' (Rose 1991), a belief that only humans have rights and that our primary allegiance (and political struggles) should be towards our own. This may be accompanied by a recognition that we might have duties towards animals – not to cause them suffering, for example – but these in no way can be balanced against human rights. This is, of course, the argument that underlines the use of animals by human societies; their use in scientific research, for example, is justified by the argument that humans may benefit. Publicity produced, for example, by the Research Defence Society emphasizes medical advances (such as the discovery that insulin can be used to treat diabetes), stressing that such discoveries were made using animals.

Throughout these various ways of thinking about humans, animals and the feminist critiques of science run two tensions. The first is the one with which I began – similarity versus difference. We think of ourselves as similar, for example, when we use animals as metaphors, as models of human society. It is similarity (or at least common ground) that is inferred in much ecofeminist writing. By contrast, the assumption of fundamental difference underpins the use of animals as objects of inquiry, that justifies their use in the laboratory. It is also the notion of difference that underpins the denial of animals in the realm of human sociality.

A related (but perhaps less obvious) tension centres on the assumptions we make about our own behaviour. Typically, feminists have rejected any notions of gender being fixed or determined, stressing instead the diversity and flexibility of social responses. One reason why (some) feminist writing which links women to the environment has been criticized is that the link is said to essentialize women, to reduce them to a fixed and unvarying essence.

This may be true of some writing in this genre, but it need not be universally so. To accept some similarity to other species is not necessarily to reduce or demean women (or humans more generally). Nor should we see nature as inherently fixed, while we take on ourselves the mantles of free will and social construction. Both, I believe, are misleading.

We need to question why we become concerned about the juxtaposition between humans and animals. The worry is always that those humans are then being relegated to an inferior, and biologically driven, category. As such, they lose those characteristics we define (and cherish) as human – rationality and free will, for example. But this move relies on a particular construction of 'the animal' – one to which science tends to subscribe. Are *all* animals really like that? Or might another strategy be to challenge the kind of thinking that relegates all other species to this lowly level of bestiality, while simultaneously elevating ourselves?

This second strategy comprises a major theme of this book. The first part explores the meanings of our ideas of 'animals', both in science itself, and in the wider culture, and how these meanings relate to our relationships with animals/nature. In some areas of feminist writing, animals appear as fellow-sufferers (in some critiques of reproductive and genetic technologies, for example). In other areas, what is done to animals is more centrally a source of concern – feminist writing about the environment, or focusing on anti-vivisection, for example. In yet other writing, animal societies (especially primates) are metaphors; here, we can ask what function does 'the animal' serve in these discourses? What does it tell us about feminism and science?

The second part of the book focuses on one specific aspect of the relationship between science and animals, namely, the use made of animals in the laboratory. I am not concerned here with listing the atrocities in the style of anti-vivisectionist literature, nor of defending the progress of science as emphasized by the scientific community (I do not want to add to the mountain of literature on either side of this polarity). Rather, my concern has been to look at the place of animals in the activities of science, and in how it is written.

The remainder of the book analyses in more detail why feminist theory needs to look more critically at its implicit reliance on particular notions of 'animal', and at what that means for how feminists think about what constitutes animality – or humanity. The context of this discussion is in the critiques of science that feminists have mounted; but it does, I will argue, raise wider issues within feminist thinking about gender. The particular focus here will be the relationship between the human versus animal polarity, and feminist critiques of biological determinism. The relationship, I will argue, is problematic and raises many questions about the assumptions that underpin at least some feminist theorizing.

My central theme throughout the book is to explore the meanings of 'animal' and 'human' in relation to the feminist literature on science. These meanings do, of course, have much wider currency, both within feminist writing and in other discourses (they relate to debates about animal rights, and about environmental issues, for example). But my aim here is to bring to the surface some of the hidden assumptions we make in feminist thinking about other kinds of animals, and our relationship with them.

As I have said, my main concern is with the *idea* of 'animals' and the way it is counterposed to 'human', but it is also my concern *for* real animals that motivates me. I do not believe that animals are something apart from human society; companion animals are undoubtedly taking part in human culture – and even wild animals must be affected by our activities.

Moreover, I care passionately about the despoliation of environments that threaten wild creatures; I care passionately about the cruelties that are daily inflicted on animals as pets, in the slaughterhouses, or the various other ways in which we come into proximity to animals. Science is not unique in the way that animals are subjected to potential pain and suffering; indeed, considerably more creatures suffer pain and anguish in being transported under appalling conditions and then squashed into holding pens before a slaughter that is sometimes less than humane. Some individual scientists may respect animals; perhaps some slaughterhouse workers do too. But on the whole, the practice of science is not respectful. That lack of respect is a stance with which I profoundly disagree, and it is deeply rooted in our cultural beliefs that we are somehow different from – superior to – the mindless morons we construct as 'animals'.

2

THE MEANINGS OF
ANIMALS

Most of them accepted namelessness with the perfect indifference
with which they had so long accepted and ignored their names.
Whales and dolphins, seals and sea otters consented with particular
grace and alacrity, sliding into anonymity as into their element . . .
None were left now to unname, and yet how close I felt to them . . .
They seemed far closer than when their names had stood between
myself and them like a clear barrier: so close that my fear of them
and their fear of me became one same fear.

(LeGuin 1987a)

How we see – and name – other kinds of animals matters a great deal
to how we think about ourselves. Not only have we come to rely historically on at least some species (for their labour or their flesh for
example), but we have also seen animals as reflections of ourselves. It
is not their views of their world that matters here, their self naming, but
how we choose to name them – as friend or foe, as similar or different.
In this chapter, I want to examine some of the ways in which we
culturally use particular ideas of 'animals', as well as looking at three
'case studies' of feminist thinking to see how 'animals' appear there.

A strong theme in feminist writing is to see animals (or nature more
broadly) as 'fellow sufferers'. Women, like animals, have been subjected
to domestication of their 'wildness', to breeding programmes, to experimental regimes, to vivisection. What these connections play on is the
recurrent association in much feminist writing between women and
nature; women are seen to be more nurturant, closer to the earth.

Feminists have also criticized science, emphasizing its power in our
society and the various ways in which it is fundamentally political; those

who do it cannot help but incorporate social values and ideologies into their laboratory or field practice, into their hypothesizing, or into the ways in which they disseminate their results. They make, as Donna Haraway reminds us, stories that tell us much about the political currents of late twentieth-century western society (Haraway 1990). Among other things, that society continues to devalue women. The stories that science tells about animals, similarly, tend to devalue them; they become 'models' for medical experiments, their sentience or intelligence downplayed.

What do we mean by 'animals'? Some meanings and metaphors

To begin with, the word 'animal' can mean many things, and can be used in many different ways (as, indeed, can the word 'nature'); it may signify a standard against which we set ourselves (in which we are better). To say, for example, that football supporters who go around fighting and wrecking places are behaving 'worse than animals' is to imply that they are out of control, behaving in a *sub*human way.

'Animal' may thus mean something in human nature that we dislike – 'animal'-like behaviour, the 'beast within'. Images of animals (typically pejorative and infantilizing images) are also invoked to describe women: I may thus become a chick, a bunny, a pussy (or, when I'm answering back, a bitch – or shrewish). These words, of course, are intended to denigrate women, to reduce us 'to the level of beasts'. What is invoked is invariably a hierarchy: men above women, women above animals/ nature. Just as men sometimes seem to dislike or fear being branded as 'feminine' because it appears to demote them, so women have come to reject any suggestion that we are close to animals.

There are many ways in which our culture uses concepts of animals; I want to explore just a few here to try to unravel a few of the meanings they evoke in the wider culture before turning my attention to science.

The beast

The beast is a powerful cultural icon. Interestingly, the *Oxford English Dictionary* distinguishes an obsolete concept of 'beast' that included humans from the more familiar form of 'later usage' that separates us from other species. The term beast has, historically, acquired many layers of meaning. Its meanings have included animals used for work or kept as stock animals (hence, beasts of burden). Another meaning of beast became associated with the devil, implying the forces of evil. When we use it today, however, it typically carries two implications: that it is pejorative, and that it is used to separate us from other species – even when it is used about other people.

Many kinds of animals have come to symbolize the evil beast, as a force associated with evil or whatever it is we dislike. The serpent in the biblical tale of the garden of Eden is one example; the roaring, man-eating tiger is another. The beast is often fearsome, or out of control; it may be the 'beast within', an association with whatever we disapprove in our own natures, such as human aggression (Midgley 1978) or something we project onto nature outside. In this case, we might invent the form to fit – the mythical dragon is an example.

'Beasts' clearly have considerable rhetorical power in the English language. They symbolize our denial of aspects of ourselves that we don't like; perhaps for that reason the notion of 'beast' seems to me to be ambiguously gendered. The icon that springs to mind when I think of 'beast' is of an aggressive, ferocious creature, maddened by some unnamed threat. Such cultural images are usually seen as male. But 'beast' also represents the denied bit of ourselves, a bit that we cannot control; for men, I suspect, that is most likely to be denial of femininity. 'Beasts' can be both.

They also symbolize western cultural hegemony over the rest of the world, its nature and its peoples; they symbolize the separation that this implies between us and other species (mere beasts); they symbolize the way that we have come to dominate nature. Beasts, however, may bear little relationship to the real lives of the animal whose image is invoked; wolves, for instance, conjure up images of beastly ferociousness. But real wolves are not the brutal killers of Little Red Riding Hood fame; on the contrary, they lead complex social lives, sharing the responsibility of caring for pups.

The evil beast of folklore is clearly a myth, yet it remains one that we project onto untamed nature, onto many non-humans, and onto many humans too. It sits, snarling ominously, alongside our attempts to construct images of animals as tame representatives of 'civilized' society. Thus we convert wild cattle into docile oxen; powerful stallions into less intransigent geldings; wolves into lapdogs. Literally, these constructions come to have a place in our society as much as we ourselves do; as metaphors, they represent the taming of the other, of wild nature. No wonder they are so loaded with cultural meanings.

The wild versus tame

A second area of animal meanings centres on our ideas of wilderness and the wild. Some animals signify for us something untameable, unapproachable by modern civilization – the soaring eagle, for example, or the call of the dolphin. People's fascination with birds seems to be part of this: most birds can fly, seemingly independent of human existence. Other animals symbolize fierceness in their wild state – the lion, for example.

The 'conquest' of wilderness has, of course, largely been the preserve of men. Until recently, only a few women were associated with the risks and thrills of wilderness exploration (Isabella Bird's accounts of her treks through the Rockies in the mid 19th century are one example). Climbing mountains or crossing gorges, pushing back the frontier, facing extreme weather conditions, and confronting fierce wild animals were long held to be activities that only men could and should do. Facing an adversary such as an angry grizzly bear is undoubtedly terrifying and dangerous, but the odds were against the bear – not because only men possessed the strength or cunning to outwit such a large creature, but because they were likely to have guns.

Wild animals were (and are) often killed to demonstrate the prowess of the hunter: the beautiful bear becomes the humble bear-rug on the floor. Some, however, are chased and trapped to become live exhibits, their wildness gone. Zoos, circuses and the like are places where we can see 'wild' animals performing for us.

We have subdued animals to make them domesticated, tamed by contrast to those existing 'in the wild'. We domesticate them as beasts of burden in agriculture, as companions in our homes, for sport of various kinds. Taming may, in other circumstances, imply removing the animal symbolically from its wildness; at its extreme, this may mean that the animal either becomes, or is controlled by, a machine. Animals have long been used as sources of power by humans – as turnspits, towing canal barges, working mill-wheels, pulling ploughs or carts. No wildness here: the 'good' draft animal is one that works steadily, without complaint, for long hours.

Taming, too, is ambiguous. It can be synonymous with training animals to carry out particular tasks; it can also imply conquering the wildness in the animal. To 'tame' a horse – a symbol of power in our culture – is to subdue its wildness, to make it accede to our wishes. Colloquially, 'tame' also means boring, unexciting ('that was pretty tame, wasn't it?'), the antithesis of wild ('that was a wild party').

In transforming animals from 'wild' to 'tame', humans reconstruct the boundary between the animal and ourselves. No longer symbolic or representative of recalcitrant nature, 'taming' recasts the animal, bringing it closer to our idea of humanity. It also brings us closer to nature, but to a nature more amenable to our control, thus blurring, a little, the human/animal boundaries. The transformation of large animals such as horses from wild to tame is perhaps particularly salient in this respect, as even 'tamed' horses have come to symbolize a power and wildness of nature that is accessible to us (see Lawrence, 1985: 195). In that sense, the idea of 'the horse' contains not only the notion that horses come to symbolize certain aspects/values of human culture, but also the notion that humans can partake of wildness and animality through our association with them.

Pests and pets

Some kinds of animals we classify as pests, as vermin. These are animals whose wildness we seek to eliminate by eliminating the animal itself. Thus, we kill locusts, tsetse flies, mosquitoes and rats. Yet it is not the *kind* or species that we seek to eliminate – just those of that kind that we class as pests. Thus we kill rats who live in chicken coops, we are exhorted to kill rats in towns because of the threat to human health. Yet simultaneously, we breed rats for laboratories, or even as pets. Foxes, too, are classed as vermin by those who defend their right to hunt them in the British countryside. In the meantime, many people living in the urban areas where fox populations are growing rapidly work to protect them.

Pet-keeping, too, is widespread throughout human cultures (Serpell 1986), but what counts as a 'pet' depends upon who you are and where you live. Some cultures, for instance, would look askance at the way my dogs live in my house and share my space as pets. In Europe and North America, moreover, pet-keeping has developed strongly only in the last two centuries. Until relatively recently, what pet owners were likely to have in common was wealth and rank. Not only did this reflect their financial status (having an animal that did not earn its keep), but pet owners may

> have enjoyed a metaphorical security – a feeling of supremacy over nature – that was as unusual as their exalted social position . . . Pet owners probably saw the non-human world as a less threatening and more comfortable place than did most of their contemporaries, who understood their relationship with the forces of nature primarily as a struggle for survival.
>
> (Serpell 1988: 19–20)

'Pets' thus symbolize human mastery over wild nature, and our power to incorporate at least some animal kinds into our social world. They also have gender connotations. Dogs tend to be thought of as 'he', while cats become 'she', for example. The gendered symbolism can also extend to their human partners: it is somehow effeminate to have a toy poodle, while Rottweilers and pit bulls seem the epitome of masculine aggression.

Meat

We have also conquered the wildness of some kinds of animals by cultivating them, and then killing them. Eating meat is an activity loaded with symbolism. The word itself has many layers of meaning: 'meat' can mean parts of the body, as feminists have often pointed out when women are 'treated as meat'. It can also mean having substance – as in the 'meat of an argument'.

Both symbolically and historically, the production of meat has involved environmental destruction; meat has, moreover, become associated with power (Rifkin 1992; Fiddes 1992). It is, as Fiddes points out, ironic that we have subdued the 'wildness' of cattle or sheep, we have rendered them docile by taming, selective breeding, castration. Killing them symbolically emphasizes their subjugation. Yet their flesh, when eaten, is supposed to give muscular strength, not docility.

Meat-eating has also been culturally associated with gender: it is manly to eat meat (especially beef) while vegetarians are effeminate. 'Real men' tame and kill other creatures: then they can become real men all over again by eating and thus incorporating into themselves the dead body of the tamed and docile animal they have killed. But the realities of that process are obscured by the language of denial in which meat-eating is couched. 'We see ourselves', notes Carol Adams in her analysis of meateating and gender, 'as eating pork chops, hamburger, sirloins, and so on, rather than 43 pigs, 3 lambs, 11 cows, 4 "veal" calves, 1107 chickens, 45 turkeys, and 861 fishes that the average American eats in a lifetime' (1990: 67).

'M/eat to live' suggests a poster I pass regularly advertising the nutritional benefits of meat. It depicts a healthy looking male runner, whose health and strength we must attribute to meat. You need real protein, I have often been told by those concerned that my vegetarian diet is insufficient to keep me going. The story constructed here is one of scientific support for being carnivorous, the properly balanced diet. No matter that protein is protein; animal protein gets labelled as superior protein. But the real meanings of meat are the ones that Adams has outlined: whatever euphemisms we choose to use, meat means dead animals.

There are obviously many other ways in which different meanings of animals may be constructed, from wild beasts to tame, or even boring, pets. These few will suffice to illustrate the breadth – and contradictoriness – of what animals mean to us in western culture. In some ways, to think of animals as natural kinds, as something inherently *in* nature, is facile; whatever animals are to themselves, we construct our own meanings and classify non-human animals in a variety of different ways. And many kinds of animals are just as much *in* human society (Benton 1993). With those thoughts in mind, I want to sketch out some of the ways in which feminists have viewed animals.

'Animals' in feminist writing

> Civilized Man says: I am Self, I am Master, all the rest is Other – outside, below, underneath, subservient. I own, I use, I explore, I exploit, I control. What I do is what matters. What I want is what matter is for. I am that I am, and the rest is women and the wilderness, to be used as I see fit.
>
> (LeGuin 1989)

> there is no shame when one is foolish with a tree   No bird ever called me crazy   No rock scorns me as a whore   The earth means exactly what it says   The wind is without flattery or lust Greed is balanced by the hunger of all   So I embrace anew, as my childhood spirit did, the whispers of a world without words
>
> (Chrystos 1983)

One strong theme in, for example, feminist fiction is a feeling of *empathy* with animals or nature, partly as fellow sufferers at the hands of an uncaring society, and partly as sharing qualities – caring and nurturance in particular. Or animals/nature may be portrayed as possessing qualities that even women might envy – 'the earth means exactly what it says', notes Chrystos in her poem. That sense of common cause with animals is strong in feminist communities; many of us live with animals and/or are vegetarian.

Here, I want to look at three areas of feminist nonfiction writing that speak of animals, relying on women's closeness to nature, to examine the meanings of 'animals' that are invoked. Each bears some relationship to science. The first concerns women's closeness to nature, evinced for instance in ecofeminist writing; the second looks at stories of animal societies; and the third, at the issue of vivisection, to which some feminists are adamantly opposed. In each case, it is similarity in the way that animals and women are treated or described that is emphasized.

Parallel lives: The taming of women and nature

One source of parallels in feminist writing has focused upon seeing women/nature/other animals as being similarly 'tamed', broken in, domesticated. By contrast, the 'wildness' of women is celebrated in some parts of feminist culture – in songs, for example, and fiction. Mary Daly invoked women's wildness in her *Gyn/Ecology*, as a means by which women might fight back against their exploitation. She urges us to 'follow the call of the wild', which means 'living in a state of nature . . . not tamed or domesticated', or 'not living near or associated with man'. It also means 'not amenable to control, restraint or domestication: UNRULY, UNGOVERNABLE, RECKLESS' (Daly 1979: 343).

There is a contrast in some feminist writing between wild nature and what happens to women or to non-white communities. It is evoked, for example, by the title of Andree Collard's book, *The Rape of the Wild*. (Collard 1988). Her book is an impassioned plea for women to listen to 'mother nature', to be at one with her, to pay attention to the violence perpetrated in our culture (largely by men) in, for example, hunting, animal experimentation, genetic engineering or the destruction of the rainforest. For Collard (1988: 137),

> The identity and destiny of woman and nature are merged. Accordingly, feminist values and principles directed towards ending the oppression of women are inextricably linked to ecological values and principles directed towards ending the oppression of nature.

The links forged in ecofeminist writing between women and nature are partly symbolic (the metaphorical associations of mother nature, for example); they are partly spiritual, celebrating communion with the earth and nature (Ruether 1989; Spretnak 1989, 1990; Starhawk 1990). They undoubtedly appeal to something deep in many of us; it is often awe-inspiring to come across a wild animal, deep in the woods, or to stand upon a mountain top high above the tree line.

Women, moreover, have much to say about the way that global ecosystems are being wrecked by human (patriarchal) greed. Insofar as both nature and women stand largely outside global systems of domination, then women generally *are* connected to nature in varied ways (see Diamond and Orenstein, 1990 for various perspectives on women and nature). Undoubtedly, different women bring different perspectives. I am white, middle-class, a lesbian – all of which mean that what I say or do about 'nature' or animals must be markedly different from that of (say) women working in urban factories in southeast Asia, or women confronting the typhoons that have swept Bangladesh. I do not, for example, have to subsist on 'nature' at all. Moreover, I live in a country where agricultural practices have shaped the landscape for millennia (and industrial practices for two centuries) – all of which shape the perceptions of inhabitants. 'Nature' here is relatively benign, even tame; the lives of people in lowland England are not usually threatened by catastrophes such as typhoons or earthquakes. A vision of a benign, tamed nature does, in some ways, seem more appropriate in the lands of northern Europe than in many other parts of the world.

Vandana Shiva has emphasized how the destruction of the global environment is tied to the oppression of women in poor rural areas, particularly in developing countries. She emphasizes that new social movements among women in such countries show that women are not the passive victims that some accounts of environmentalism and feminism might seem to imply. On the contrary, 'women and nature are associated *not in passivity but in creativity and in the maintenance of life.*' (Shiva 1989; emphasis in original). She notes that,

> in the perspective of women engaged in survival struggles which are, simultaneously, struggles for the protection of nature, women and nature are intimately related, and their domination and liberation similarly linked.

(Shiva 1989: 47)

The strength of ecofeminism is its insistence that women have something to say about nature or animals. It is, quite rightly, generally critical of science, as well as of the wider environmental movement. Certainly feminists have been critical of the 'male ecology movement' which often seems to valorize 'mother nature' while at the same time ignoring women's interests (Doubiago 1989).

Nor are the interests of (at least some) animals well served in environmentalist concerns. Interesting (to some humans), endangered, or otherwise exotic kinds of animals may be protected, at least as populations. Individuals rarely are, however, and it is tough luck if you happen to be an individual of a species that someone, somewhere, thinks is getting in the way of the wilderness. As I write, wolves are being massacred in the 'wildernesses' of Alaska, for just such reasons. The argument is that they are being 'culled' (a word which sounds somehow more scientifically, environmentally, respectable than 'killed'; anyhow, you cull populations, but it is individuals that get killed). The culling, we are told, is necessary because they are beginning to decimate the caribou herds. Well certainly we may wish to protect the caribou, but why are *they* being protected? So that human hunters have plenty to shoot at. The animals, in this tale, become either a 'protected resource' (if caribou), or a threat to that resource and to human exploitation of it (if wolves). That both caribou and wolves might have an interest in all this seems not to be part of the equation.

Ecofeminist writing, of course, explicitly addresses women's issues (and is unlikely to support the wolf shoots). The problems I see in ecofeminism are not to do with ignoring women or our perspectives, but of a tendency to see both women and nature as benign, pacific, ultimately good – the antithesis of the exploitative and violent values that are helping to destroy us all. But this is misleading; nature is not always benign, any more than women are. Indeed, as Mills (1991: 167) points out, the trouble with a view of nature as benign is that it ignores the reality that ' "Nature" in its "wisdom" creates not only ecological balance but unwanted pregnancies'. Nature – like women – can be capricious.

If nature is thought of as benign, then so too can the animals that are part of it. Think again of wolves. We can reject the fierce beast of mythology. Or, we can paint a picture of these creatures as a kind of loving parent. This, for example, is what Lois Crisler did in her accounts in the 1950s of living with wolves; the stories she told spoke of wolves 'smiling', and of their commitment to family. She spoke of how these animals could 'read your true feelings' from your eyes (Norwood 1993: 242), even of their sense of 'social responsibility' (Crisler 1991). But those claims, too, are inevitably stories. Crisler wanted to stress the 'family life' of the wolves for a film documentary; perhaps in the 1990s she would have told another tale.

The picture of nature in ecofeminism is largely benign, the fertile earth mother. Animals, as part of nature, are fellow sufferers. While that may be true to some extent, we need to be wary of telling tales that simply cast animals (or women) as always good, for it is surely as much a metaphor as the evil beast.

Mirrors of nature

A second way in which animals appear in feminist writing is in accounts of animal societies. The most familiar use of animal parallels is, of course, in the kinds of accounts that feminists have criticized. We are familiar enough with the kind of argument that runs: look at other animals – baboons, chimpanzees, large carnivores – and see how dominant the males are (for critiques from a feminist perspective see Bleier 1984; Hubbard 1990). Sometimes we have looked at the animals and at what has been written (by largely male scientists) and then criticized the content of the writing for its biases. But at other times, women have written alternative accounts, telling a more 'female', or feminist, story.

One reason why animal societies (particularly primates) have been so assiduously studied rests on the belief that to understand society we must understand our 'animal nature'. This belief in the animal within rests uneasily alongside cultural distinctions between human and animal. It is the mythical 'animal nature', at the core of human nature, that has been subjected to the forces of evolution. Human culture, according to this view, is not only the icing on the surface, but can be best understood in terms of the underlying cake.

The trouble is that it is indeed a mythical animal nature at the core. Donna Haraway (1989b) has documented how primatologists in the 1940s saw pictures of male domination and competitive societies, while later accounts by women primatologists influenced by the women's movement told tales that centred on the females in the group. Different observers tell different stories, and come to those stories from particular cultural backgrounds.

Some writers have aimed to 'redress the balance', sometimes sensationally. One such book, for example, claimed to uncover the important role of females in nature, speaking enthusiastically of 'Liberated Ladies of the Animal Kingdom' – species of animals that defy the 'coy female' stereotype (Shaw and Darling 1985). Yet it is easy to find examples of species that 'don't fit' a particular pattern, for it is indeed our own biases that assume that one pattern is somehow more natural than others. Whatever the writers' motives for writing about female animals in such detail (and they do claim that it was written in response to sexist biology), the book is a clear example of how powerfully language constructs our images: here, the language of 'women's lib' (*sic*; this is not the language of feminism, but the language of newspaper parodies of feminism) seems inappropriate to the worlds of animals. Liberated ladies? Wasps that raise their macho flags? Stickleback fathers who get custody? This is still an account that writes gender stereotypes onto other kinds of animals.

More circumspect examples outlining the importance of females in animal societies are careful to avoid such colourful language (although they retain the positivist language of supposedly factual statements). Irene Elia, for example, takes care to remind the reader that 'Nature does not restrict herself to manmade categories. The taxonomic scheme may be used to organize data about females in a general way, but the categories may never be considered absolute' (1985: 8).

What such texts as these indicate is an attempt on the part of women to create what Donna Haraway has called another lineage. As she points out, referring to one of the male proponents of 'popular' sociobiology, 'Barash's concern for lineages is his central rhetorical strategy. He calls on the authority of the father and names it scientific knowledge' (Haraway 1991: 73). Putting female animals into an alternative lineage goes some way toward creating an alternative tale. Elsewhere, Haraway also notes that 'There seems a grain of truth . . . that simply putting females at an explanatory centre is in some sense feminist.' But, she reminds us, 'not just any story will do'; primate stories are contested, and one source of that contestation has been contemporary political struggles over women's roles (1991a: 107).

So in these stories of animal societies, we can seek mirrors. They construct images of animals as just like us: whether the storyteller is avowedly antifeminist or creating alternative tales, the female creatures populating the stories conveniently either hold the apron strings or wave the feminist banner. What they have in common, of course, is that they are constructed as *scientific* tales. We are meant to believe in these accounts as scientific truth; it is this 'truth' that gives them powerful meaning.

This is what animals *really* do. But the scientific voice merely hides their role as metaphors, particularly metaphors of western culture and society. Where is the wolfness of the wolf in the images of happy families?

Animals in the laboratory

Although less obviously connected today, feminism has in the past been concerned with the fate of animals in laboratories. Many of our forebears in the late nineteenth century saw links between the fate of animals in science and the possible fate of women. In general, awareness of animal cruelty grew during that century (Brown 1978; Ryder 1989), alongside a growing concern about what specifically went on in the labs. By the 1870s, concern was sufficient for a bill to be put before the British Parliament to control the use of living animals in experiments. Inevitably contested by the scientific community, the bill appeared in a less stringent version as the 1876 Cruelty to Animals Act (now superseded by the 1986 Animals (Scientific Procedures) Act).

At about the same time, popular opposition to vivisection was growing and it had strong links to feminism. Frances Power Cobbe was an ardent supporter of the Victorian anti-vivisection movement as well as an outspoken feminist. Anti-vivisection was the cornerstone of late nineteenth-century radical criticism of science, evoked in response to what was perceived as the threat of the growing power of the medical profession. Science was seen as meddling dangerously in nature and health. In this perception, the anti-vivisectionists were allied with movements opposing vaccination and the implementation of the Contagious Diseases Acts (Elston 1987).

The reaction of scientists was equally strong, occasionally making specific reference to 'womanly sentimentality'. In a letter to a magazine in 1875, for example, one male scientist defended his work against opponents on similar grounds, claiming that fishing was far more cruel, and that scientists were the best judges of their own work; the letter illustrates not only the defence of animal use, but the use of language that would be abhorred today.

> I can positively say that the fisherman caused more suffering in one afternoon than I have caused by all the cats put together that I have skinned alive all my life. Some thousands have passed through my hands in my time. The baits he used were hours in their torture while an awkward fellow even can strip a cat as clean as your hand in from 10 minutes to one quarter of an hour from the first touch of the knife. In my own case, when the skin is once fairly off, as often as not I put the poor things out of their pain at once. Why shouldn't I? I never wish to cause unnecessary suffering only like

my brother vivisectionists I claim to be the best judge of what is necessary, and like them I don't wish to be interfered with by whining sentimentalists.

(Brown 1875: 817)

Earlier in the letter he had referred to 'weak-minded if well-meaning women and men, still weaker than women'.

'If vivisecting indicated depravity in a man, in a woman it would be even more horrendous', Elston has pointed out (1987: 281); much concern was expressed publicly around the turn of the century about the effect of such practices on women's sensibilities. The concern from the anti-vivisectionist side was matched in the writing from the defenders of the practice. The quarterly report of the Research Defence Society (founded in 1908) of October 1919 was concerned to reassure would-be women scientists that animal experimentation could be suitable work for women:

This sort of work, like holy matrimony, is 'an honourable estate' not for men only, but also for women; and last year, 45 women were licensees under the Act . . . Women can make themselves quite as skilful as men at all the delicate methods of practical bacteriology and practical pharmacy; and the standardising of antityphoid vaccine is work just as philanthropic, and as womanly, as the nursing of cases of typhoid.

(Research Defence Society 1919: 2–3)

It was, however, precisely the entry of women into physiological research that challenged women's dominant position within the anti-vivisection movement. The movement in turn reacted by emphasizing the degrading effects of such cruelty on women themselves. 'A movement that began by claiming women's special tenderness and kindness was, twenty years later, warning against women's capacity for evil and cruelty', notes Mary Ann Elston (1987: 286). Part of the problem for anti-vivisectionism, she suggests, was that it had appealed to 'natural' differences between women and men, to women's 'natural' capacity for kindness; but the alternative, equal rights, view was gaining credibility (Rosenberg 1982; Elston 1987). Entering the professions on an equal footing to men was increasingly part of the feminist agenda.

Within ecofeminist writing, a concern for the fate of animals in laboratories has returned to the feminist agenda. Groups such as Feminists for Animal Rights in the USA make explicit links between feminist politics and animal rights, for example, while feminist anthologies of ecofeminist writing sometimes include articles specifically condemning vivisection. The link, as it was in the nineteenth century, is that both animals and women are seen to be suffering at the hands of science. Norma Benney, for example, reminds her readers that feminism's central concern is overthrowing the patriarchy, but we should not ignore animal suffering in doing so. 'If we struggle to free ourselves, without realising that we are also crushing the most oppressed and exploited creatures on the planet, we can only fail', she believes (Benney 1983).

Here, as in ecofeminist writing more generally, 'animals' are fellow sufferers. The animals that populate anti-vivisectionist writing are almost invariably mammals or birds, usually cute, cuddly ones with which the reader can identify. Mammals, of course, have nervous systems like ours, and no doubt do suffer and feel pain as we do. And appealing to our identification with furry bunnies or monkey mothers is no doubt effective as a campaigning strategy. The sense of revulsion at seeing some of the lurid photographs of suffering animals certainly got to me early in my life; I recall at junior school trying to recruit my friends into an anti-vivisectionist organization after seeing pictures of rabbits with sore eyes. In the 1950s, of course, animal rights was not a popular cause, and I don't remember being particularly successful!

My concern in this chapter, then, has been to sketch some of the different meanings we attach to concepts of 'animals'. Some words in English carry powerful connotations, 'beast' and 'wild', for example. Some, like 'meat', seem to gloss over the animal realities. Where feminist writing has dealt with animals, we have generally invoked a sense of animals as fellow sufferers at the hands of a rather cruel society. That may indeed be true in many ways; animals are often treated as nothing but commodities just as women sometimes are. But we should be careful of seeing 'animals' as a kind of homogenized mush. The meanings we attach to them can be highly variable; we need to bear that in mind in thinking about how science writes about animals. Scientific meanings are just as contested as those we use in the wider culture. So, if we are going to think about, say, the way that animals are treated in science, then we have to think about how different meanings of 'animals' are used in different contexts. In the next chapter, I want to sketch out some of the ways in which the meanings humans have given to 'animals' in western culture may have changed historically, before turning my attention in Part II to animals in science.

3

CHANGING MEANINGS

[Holsteins are] allowed to live only $4^{1}/_{2}$ years . . . they are industrial animals, no more fuel efficient than a mammoth tractor, suited to the richness of an industrial economy.

(Klinkenborg 1993)

Here I want to explore how our ideas about animals and their place in nature have changed historically, alongside social changes, including beliefs about gender. How, for example, have social and political contexts – such as the processes of industrialization – helped to shape the *meanings* we give both to the concept 'animal' and to specific kinds of animals? One change, illustrated in the statement about Holstein cattle, has been towards creating commodities out of certain kinds of animals. These changes provide a backdrop for later chapters in which I explore how ideas about animals have developed in some areas of scientific thinking and writing. Such developments in science cannot, however, be separated from the ways in which animals and humans are thought about within the wider society.

As I have noted, ecofeminist writing in particular focuses on belief in an affinity, or closeness, of women to nature. According to such views, animals are part of nature and women are closer to animals – an association explored in much feminist fiction too. By implication, men are associated with culture. That the duality between nature and culture is linked with gender in western thought is a theme explored in much feminist writing; so, too, has the ways in which that association has changed over time. The dichotomy, and its apparent rigid association with gender developed particularly in Victorian times; prior to that, more flexible notions of gender, and of the relationship of nature to culture, seemed to be evident (Jordanova 1991).

Locating some people on the side of 'nature' has had important political consequences, and it seems no accident that these beliefs developed when they did. In the nineteenth century particularly, at a time of rapid European expansion and imperialism, it was undoubtedly useful to be able to cast women and 'inferior' races as closer to nature. But it was, of course, not a peaceful, sylvan image of nature that was invoked in these associations (so unlike the benign images of mother earth that pepper much environmentalist writing today). Rather, the concept of nature that tended to find its way into accounts of non-white races, or non-British ethnic groups, was one of an uncontrollable, wild and predatory nature – the beast without. Thus one nineteenth-century writer spoke of 'careless, squalid' Irishmen, 'living in a pigsty . . . multiplies like rabbits or ephemera' (quoted in Jones 1980: 102). Similarly, women's sexuality was seen sometimes to be 'animal-like' – particularly if the woman was black.

So proximity to nature and the stigma of being 'like animals' were invoked as a means of justifying various forms of oppression, on the grounds of gender, class and race. Not surprisingly, then, many feminists have sought to deny women's 'animality' and struggled to be admitted to the hallowed ground of human culture, but there are many problems with this approach. One, for example, is that to deny 'animality' is to assume essentialism. We may (and feminists often seek to do so) try to avoid essentializing ideas about 'women', but the animality of 'animals' – the beast again – goes unchallenged.

Yet alongside the development of the nature/culture dichotomy, non-human animals themselves have been cast by humans into particular roles; often, these roles reflect the social status of the person to whom the animal is related or belongs. At times they may reflect the uses to which people put the animals. Those roles are always changing, however, and their meanings constantly negotiated within human cultures and social structures.

Changing perceptions: Animals and their place in nature

Popular ideas about, and relationships with, other animals have changed considerably over the centuries. So too have the meanings that we might attribute to animals. Here I want to sketch some ways in which historians have described changing attitudes towards animals, in the context of social change.

One significant change seems to have been in beliefs about the abilities or responsibilities of non-human animals. Today we would not consider that an animal could be held responsible in law for its actions (although we might expect its owner to be). Yet during the Middle Ages, non-human animals seem to have been held morally responsible, and therefore potentially culpable. Thus animals were sometimes subject to trial and possible execution as criminals; in his history of these trials, Evans (1987) notes the case of a pig in Normandy in 1386 which was tried for murder. The animal was dressed in human clothes and hanged from the gallows in full view of the town populace following its conviction.

For many centuries, the place of non-human animals in European views of nature (or, indeed, of human animals) was seen largely as given by God; the Bible, for instance, refers to human *dominion* over animals as desired by the deity. Two important sets of ideas structured how western people tended to see their place in nature, at least until the eighteenth century. One of these was the idea of the 'chain of being' which conceptualized humans as superior to animals (Lovejoy 1936); the other was the notion that non-human animals could be classified according to the uses to which people put them and/or according to their 'characters'.

The dominant notion, that humans were intrinsically superior to nature/animals, began slowly to change after the seventeenth century (Thomas 1983: 301). During this period, suggests Thomas, there emerged two apparently contradictory beliefs. One was a belief in the radical subordination of nature to human intervention (contributing to nature's exploitation); the other emphasized the interconnections of nature and humanity (giving rise eventually, for example, to notions of evolutionary change and similarity).

These changes in the dominant ideology helped to change meanings. Thomas notes how legal definitions of wild and tame contributed to popular unrest during the seventeenth century. Up until then, wild animals had been defined in English law as *not* being private property, unless they were dead or tamed. Agricultural animals, by contrast, were given meanings through their status as property (chattel comes from the same etymological root as cattle, for instance). The poor, suggests Thomas, 'accepted the private ownership of domestic animals, but they held to the old common-law view that there was no property in wild ones' – the killing of which should therefore be available to everyone (1983: 49) – but access to wild animals was becoming increasingly difficult as the land was enclosed to become private property.

Attitudes towards cruelty to animals also changed in a context defined in Britain by social class. Historian Harriet Ritvo notes, for example, how acts of cruelty towards animals came to be seen as a form of social deviance, particularly if they were carried out by working-class people (Ritvo 1987). Concern about cruelty to animals was, initially, the prerogative of the richer classes.

Some early attempts to legislate were opposed in Parliament; William Windham blocked such an attempt in 1800, for example, on the grounds that the rich should not interfere with the sports of the poor. The first anti-cruelty legislation in Britain was brought in 1822 by Richard Martin, owner of vast estates in Galway. Martin was said to be a flamboyant character, who had a reputation for consideration both to fellow humans as well as to other animals. In practice, however, it was Martin himself who brought the first prosecution, against a costermonger, Bill Burns, who Martin claimed, had been beating his donkey. Showman that he was, Martin brought the donkey into the court to reveal its wounds – to the mirth of the crowds (Ryder 1989). While the meanings of animals to humans were being contested in the wider society, scientists and naturalists were busy creating another set of meanings through the processes of naming and describing. Colonial expansion meant not only the discovery of new species (by Europeans); it also entailed the development of systems of classification that were themselves mired in assumptions of European superiority, and which, inevitably, paid no heed to the classification systems of peoples indigenous to the areas from which the animals were taken.

By the nineteenth century, in the heyday of Victorian empire, the names of British imperialists were frequently commemorated in the Latin names of newly discovered (or named) species; thus, Ritvo (1990) notes how the thirteenth Earl of Derby lent his name to several species such as Derby's kangaroo (*Macropus derbianus*), while Sir Stamford Raffles's name appeared in the names of some 23 species found in southeast Asia. The classifiers also sought to disparage others who might also make claims. Ritvo, in her account of the struggles of Victorian scientists to command systematics, quotes from the Nomenclature Committee of the British Association for the Advancement of Science (1841), noting the fears of the committee that 'names that instead of commemorating "persons of eminence as scientific zoologists", celebrated "persons of no scientific reputation, as curiosity dealers ... Peruvian priestesses ... or Hottentots" ' (Ritvo 1990: 7). Only white patriarchs of the British Empire could, it seems, be trusted to name nature.

Conquering wildness

The scientific naming of species, then, was deeply rooted in the ideologies of imperialism. British upper-class men went off to distant lands and brought back 'specimens' of strange creatures, giving them names in the process. Occasionally, those names might have been borrowed from the words used by local peoples (such as the now-extinct quagga of southern Africa, *Equus quagga*); more often, local names were ignored. Obtaining specimens, of course, relied on the (unacknowledged) expertise of indigenous peoples. The drive to provide western museums with specimens for science, moreover, meant that indigenous people themselves were seen as fair game; the skeletons of Australian Aborigines found their way into the display cases of London museums, for example, while the bodies of native peoples turned up in Washington and New York. In a few cases, living people were brought to the urban centres; some tragically died, their remains subject to scientific investigation (one infamous example was the father of Minik, a Greenland Inuit boy, who spent part of his life in the USA. Minik never quite recovered from the shock of finding his father's skeleton in the American Museum of Natural History; Harper 1986).

To be measured and classified, wild animals were usually killed and brought back to European laboratories. Big, fierce (and usually male) animals were sought by nineteenth-century hunters, and displayed as part of tableaux to demonstrate British colonial power (Ritvo 1987). Not only was imperialism supported by the rhetoric of big game hunting, so was sexism and racism. Whereas pig-sticking was considered a heroic sport, most suitable for the British Raj (who hunted for pleasure; any-one – such as the local people – who hunted for profit was frowned upon), deer coursing was '"fit only for the false and effeminate natives of India"' (Ritvo 1987: 268). And it was preeminently male animals whose ferocity had to be overcome.

The animals, and their deaths, thus served as metaphors of social meaning (as did the display of the bodies or remains of non-European humans). The trophies had to be displayed in suitable fashion, not only to 'impress the natives' before leaving but also to reinforce notions of imperial might at home. The rhetoric was one of heroism and overcom-ing impossible odds (in relation to both the animals killed and the local people).

The ethos of big game hunting certainly emphasized virility; but women did sometimes enter into it nonetheless, often justifying their activities as contributing to science. One such woman was Martha Maxwell (1831–81), a feminist who believed that her work in collecting and displaying animals helped to prove women's capabilities. To this end, she went out and shot large animals, including predators, before skinning and stuffing them. Among her displays was one produced for the Philadelphia Cen-tennial Exposition in 1876. But, however much she sought to challenge sexist beliefs about women and their capabilities, her tableaux still re-flected her status as a woman in that she focused specifically on *family* groupings (Norwood 1993).

Constructing animals, in the image of man

The relationship between humans and animals in nature was not only being contested in the construction of names, however. Increasingly, people were literally reshaping animals through selective breeding. Doing so has not only meant altering the characteristics of particular species, but also distributing them throughout the world into habitats in which they would not be found in the wild. Domestic chickens came, probably via central Asia, from ancestral Burmese red jungle-fowl, for example; modern sheep probably derive from the Asiatic mouflon (Russell 1986; Clutton-Brock 1987).

Animal breeding for human use in agriculture has, of course, been going on for centuries, but it was only in the eighteenth century that systematic records were kept, and animals bred for particular inheritable traits. At that time, specifically fat animals were bred, as fatty meat was prized. One example was the much-publicized Durham ox, whom many people paid to see; so corpulent and prized was he, that his owner had him transported about for public show in a specially designed carriage (Quinn 1993).

Beliefs about animal breeding during the eighteenth and nineteenth centuries were, however, shaped by more than just the empirical observations of how traits were passed on to progeny; they were also shaped by prevailing social beliefs, particularly about gender. Breeding reflected beliefs about femininity and masculinity; only the most feminine cows and masculine bulls should be selected (Ritvo 1991).

Elsewhere Ritvo (1990) describes how 'A collection of cattle portraits, the record of generations of celebrated stock, resembled a gallery of distinguished ancestors, and a collection of cattle pedigrees was like a family tree'. Quoting a contemporary observer, she notes that the example of livestock breeding in the eighteenth century, 'led men to choose their wives as they would a brood mare, with a great care for their personal genetic inheritance' (Lawrence Stone, quoted in Ritvo 1990: 60).

Animal breeding continues to echo particular social values (reminding us that domestic animals are importantly part of *human* social structures). Animal breeders still take great care of pedigrees; what seems to be at issue in the maintenance of particular breeds of dogs, for example, is the purity of the 'bloodline' (a telling phrase), rather than the welfare of the animals. Thus, outcrossing to overcome some of the genetic problems of extensive inbreeding is strongly resisted.

In one sense, of course, selective breeding reinforces human control over those species of animals – they become tame, domestic – but in another, people have bred animals to fill or to model a particular niche in human society. In that sense, the idea of the animal that is constructed is not simply as 'other', but has a much more complex relationship to human society and needs. One relevant example is the development of 'fancies' in the late nineteenth century. Various groups of breeders evolved new 'fancy' types of all kinds of animals, from dogs and cats, to guinea pigs and rats. The fancy strains were created to fill a niche determined by (human) social status and sense of aesthetics. But their development also involved considerable knowledge of practical genetics. Thus the early development of laboratory strains of rodents at the beginning of the twentieth century owed much to the knowledge of the fanciers (Wright 1922).

Dog breeding, too, has gone on for centuries, at least among the aristocracy, but it acquired new perspective in the nineteenth century. In earlier centuries, domestic animals were typically classified not by breeds but by their usefulness to humans. Dr John Caius's book, *Of English Dogges* (1576), for example, divided dogs into '"generous", used in hunting or by fine ladies; a "rustic" kind, used for necessary tasks; and a "degenerate", currish kind, used as turnspits and for other menial purposes' (Thomas 1983: 55). Within these utility categories could be found some breeds or types recognizable to us today, greyhounds and lurchers, for example, have been bred for many centuries for hunting (Lennox 1987).

Breeding animals as pets, however, allowed people to produce animals as symbols of particular status. The animals, moreover, were assumed even to observe the class barriers of their human owners (Ritvo 1987: 90). Dog breeding came thus to generate animal mimics of human (Victorian) society, complete with élite aristocrats – the well-pedigreed winners in the show ring – and useless miserable curs. In some cases, the aim of breeding had more to do with producing winners in the show ring than improving any working qualities the breed might have; collies, suggests Ritvo, were a good example of a 'breed reconstructed to express the figurative needs of fanciers . . . [by introducing] modifications and improvements, which were tested not against the rigors of the Highland winter, but in the fashionable marketplace' (Ritvo 1987: 113).

A particularly clear example of an animal constructed to fit a particular place in/model of human society is the racehorse, the development of which reflects particular ideologies not only of social class, but also of gender and race. The modern form of horse-racing developed in the eighteenth century, although efforts to breed horses specifically for speed in racing predate that (Willett 1966; Russell 1986). To have a fast horse has long been associated with high social status, and it was undoubtedly the aristocracy who were primarily involved in breeding experiments. This association of 'quality' horses and wealth was made more explicit after the Restoration, when winners could win cash prizes (Russell 1986: 94).

Not surprisingly, the racehorse has come to represent an aristocratic élite, and has been so constructed from the earliest records of selective breeding (the thoroughbred Stud Book began in 1790, although some records go back further). During the seventeenth century, several aristocratic breeders were using stallions from 'exotic' breeds from the Middle East, putting them with locally bred mares to produce faster offspring (Arab horses particularly had, and still have, a reputation for speed and stamina); the origins of contemporary racehorses are obscure, but are generally believed to lie with three or four 'foundation' stallions of Arabian type.

By the middle of the eighteenth century, textbooks were extolling the virtues of selective breeding of horses, even for farming. Interestingly, the advocates of 'exotics' were less keen when it came to breeding these 'lower class' animals; most 'spent little time on this irrelevant subject and advised their agricultural readers to stick to native English horses' (Russell 1986: 114).

If using 'exotics' was the privilege of the upper classes, however, what they were developing came to be called the *English* thoroughbred – representative not only of emerging empires but also of those who ran them. What was important in deciding the relative status of these animals was 'blood', the quality of the breeding (as indeed it was for humans, particularly in the pedigrees of the upper class; Russell 1986: 100). Even today, the appropriately-named bloodstock industry continues to place enormous emphasis on 'bloodlines' or pedigrees.

Racehorse breeding also managed to encapsulate prevailing ideologies of gender. Stallions were believed to be prepotent, 'stamping' their qualities onto their offspring, whereas mares tended to be seen (as Aristotle had described female mammals, centuries before) as contributing much less. These beliefs continue today; it is common practice in the equestrian world to cite an animal's sire, but not always its dam. The myth persists in articles about breeding; thus, de Moubray suggests that articles claim of the Derby winner Shirley Heights that 'the influence of his sire Mill Reef was able to overcome the plebian attributes of his dam Hardiemma' (de Moubray 1987: 11). Maleness, in the world of people associated with horses, reigns supreme!

The breeders' worlds are thus influenced not only by science, in their knowledge of how genetics might work, say, but also by cultural beliefs. It is not just the specifics of how genes or traits might be passed on that becomes part of that knowledge; the science sits alongside beliefs about animals that are deeply rooted in assumptions about gender and class. When I meet people who breed horses, I can recognize the influence of science in the way that they study the basics of genetics; yet I also hear them speaking of their dismay when they breed a chestnut filly. Females of that colour are stigmatized both for their sex and for their colour, alleged to be hopelessly difficult, and therefore hard to train – or sell. It seems a curious form of sexism, having no foundation in the scientific discourse on which the breeders also depend.

What people have bred, in the form of the élite racehorse, is an animal artifact – specifically, a supreme animal athlete. Not surprisingly, there is much concern about how equine speed and performance could be improved; there is, after all, an enormous amount of money to be made if your horse wins the Derby. Yet however hard breeders try, they have generally failed to improve on racing speeds for many generations, probably because the pool of genes within thoroughbred horses is quite small. Inbreeding cannot help, but perhaps physiology can; the supreme animal athlete is now intensively studied by a form of equine sports science. Its muscle power has come under the microscope (Snow *et al.* 1983).

Meanwhile, scientific developments further change the face of animal breeding. Now, as the quotation at the start of the chapter suggests, we breed cows that resemble milk machines, the industrial animal of the twentieth century, unable to stand on her legs after a certain age. Strains of chicken bred for meat likewise cannot manage to stand as fully grown adults; they are not bred to survive that long.

Even the breeding of pet animals has moved from Victorian 'fancies' into the technological society of the twentieth century. At the time of writing, a doggie sperm bank has recently started in California. Reminiscent of the human sperm bank set up in the 1970s to provide sperm from men with high IQs, the canine version (the Canine Cryobank and Animal Fertility Clinic) offers sperm from 'blue-blood dogs and cats' (such as particular sled dogs) and abortions are offered (at $350 each) in the event of fertilization from 'substandard sperm'.[1]

Creating animals along lines dictated by human whims is limited in breeding programmes by our knowledge (or lack) of just how (or if) particular traits are related to particular genes. But new developments promise, or threaten, to reduce that limitation. In the future we may be able to tinker directly with genes and create a whole new range of possibilities. Animal bodies can, for example, be used to produce materials that we can use in medicine – machines, indeed.

How we see animals and the meanings we give them are social constructions. However much science purports to study nature 'out there', and however much nature-lovers may sing the praises of communing with wildness around them, 'nature' and its constituent organisms are ideas deeply rooted in western ideology. That ideology has contributed to notions of different kinds of animals that relate to gender or social class; fierce beasts acquire the demeanour of virile masculinity, purring pussy-cats are the epitome of cuddly femininity, while shrews were for centuries synonymous with evil.

The gendered associations of animals and their changing meanings in western culture are clearly relevant to feminist thinking about nature, but the constructions of ideas of 'animals' in science have another importance too. As I discuss in later chapters of this book, scientific discourse and practice relies on a concept of (laboratory) animals as somehow rooted in biology, eternally fixed in nature. It is precisely that kind of construction that allows scientists to make claims about, say, the hormonal determination of behaviour in a rat. From there they sometimes then jump to making claims about humans, that, for instance, is the background to claims about gender and the human brain, or about homosexuality and the brain. These claims, which feminists have often contested, depend upon naming non-human animals as products entirely of their biology.

I would not go so far (as some might do) as to claim that non-human animals do not exist at all apart from our naming; I think they do have existences, and experiences, independently of us (indeed, it is precisely those existences and experiences that we have failed to respect sufficiently in science). But I would claim that we can have no knowledge of them outside of the context of our naming.

Part of that context is science, which not only studies animals but helps to name them through its classification and descriptive practices. In the next two chapters, I turn to the specificities of animal use within laboratories, to ask the question of how animals are seen to fit into the practices of science and into the narratives that emerge from the experiments – the accounts regularly published in scientific journals. What kinds of meanings about animals (and implicitly about humans) do they convey? What kinds of stories do they tell?

Note

1 Source: *Animals Agenda*, May 1993. Information obtained through Compuserve network, Forum of the Humane Society of the United States; notice posted 11 July 1993.

Part II

ANIMALS IN SCIENCE

4

INTO THE LABORATORY

In her short story, 'Mazes', Ursula LeGuin describes how an alien torments an intelligent creature, putting it into mazes, expecting it to learn how to press buttons for food rewards or to avoid punishment. The creature ponders on what motivates the alien; it

> has never once attempted to talk with me. It has been with me, watched me, touched, handled me, for days: but all its motions have been purposeful, not communicative. It is evidently a solitary creature, totally self-absorbed.
>
> This would go far to explain its cruelty.
>
> (LeGuin 1987b: 64)

The story, she points out, is not about rats. Nor is it about how humans might be treated by aliens from another world. The alien in the story is human, the trapped creature clearly a highly intelligent one from another world. LeGuin's tale powerfully reminds us of the assumptions we make in keeping other animals in laboratories.

Animals are clearly part of what science studies, both in and out of the laboratory. Scientists interested in the behaviour or ecology of a particular species usually study those animals in the field; others, interested in physiology, say, study how animal bodies work in the laboratory. Still others may use animals as part of a series of tests to ensure the safety of particular products – drugs or cosmetics, for example. These developments have meant that animals are usually purpose-bred for research, and kept in locations specifically designed for the purpose. But their use is increasingly questioned by people outside science, many of whom feel that at least some uses of animals in research are of dubious benefit. In that context (and the context of personal threats to researchers from animal rights protesters), it is hardly surprising that those involved in doing biological research are often ambivalent about it.

Science itself is, in its practice, ambivalent about the use of animals. Their use in experiments is sometimes justified on the grounds that animals can provide 'models' of human physiology; rats and mice, for instance, are thought to be sufficiently similar to us that they are used to test products such as potential drugs. Yet their difference from us is also clear in the way that scientists must justify their use of animals to a wider public; experiments cannot ethically be justified on people, the reasoning goes, but animals can be used because of their different (lesser) ethical value.

In this part of the book, I focus upon how scientists deal with the ethical dilemmas of using animals. My primary concerns here are twofold; first, I will explore some themes about the place of animals and people in scientific research. The meanings of the animal in the laboratory are located in a web of social relationships; what the animal means to the technician is different from what it means to the scientist. What it means to the anti-vivisectionists is clearly very different again. These meanings depend not only on social relationships, but also on beliefs about animals/nature, and about the ethics of using animals for human gain. In the subsequent chapter, I will turn to considering how the relationship between scientists and animals in the laboratory is represented – at how scientists deal with the meanings of animals in their conversation and in their writing.

The controversy over 'vivisection' is an old one, as we have seen. Although feminist involvement in the anti-vivisection campaigns is perhaps not as strong as it was in the late nineteenth century, there is undoubtedly concern among some feminists today for animal rights. A US group calls itself Feminists for Animal Rights, for example, and many feminists have become involved in anti-vivisection work. If feminism is concerned with issues of oppression and exploitation, the argument goes, then we should care about the exploitation of animals.

In addressing the question of how animals are treated in science, I am assuming that there can be something to say about it from a feminist perspective. I believe that there is; as I will argue, my stance is that feminist standpoints can yield particular insights. Whatever one's personal feelings about animal rights in general, or anti-vivisection in particular, the use of animals in the laboratory is problematic. In the first place, the ethical debates are grounded in assumptions about domination and superiority that feminism, and any other social movement seeking an end to oppression, should challenge. The testing of innumerable products on animals, for example, is demanded within a capitalist society;

both consumer demand and the profit margins of industry maintain that practice. It is deeply rooted in systems of domination.

Second, the place of animals in science is particularly important for feminists because it is within the ideology of science that naturalizing definitions either arise or are justified. The pursuit of the 'homosexual rat' in scientific research is a short hop away from seeking biological bases for homosexuality in people – a search that, as I write, is rapidly gaining momentum with claims for evidence of homosexual brains, or homosexuality-inducing genes on the X-chromosome. *How* animals are used in the creation of such ideology and how that practice is written about raise important questions for feminist critiques of science.

Standing in stark contrast to the unemotional and detached language of the scientific reports (to which I return in more detail later), are the emotive images and language of anti-vivisectionist writing. A report from the British Union for the Abolition of Vivisection, for example, invites you to imagine driving down a quiet English lane and coming across a building that is obviously high security:

> behind those brick walls and blanked out windows are thousand upon thousands of animals being forced to take part in toxicological research. Something you could not possibly have known from simply driving past. But then you weren't supposed to. It was here . . . that I was soon to become an employee in an attempt to uncover the secrets so carefully stowed away from public scrutiny.
>
> (Kite 1990)

The writer is present, clearly inviting the reader to ponder the atrocities within; the photographs of pathetic beagles reinforce the emotional impact.

Scientists in anti-vivisectionist literature often seem to be seen as ogres, intent on causing animals to suffer. At the very least, they seem to be divorced from any finer feelings about the possible suffering of the animals they use in their experiments. Such images occur in feminist writing too, where the emphasis is largely on the fact that scientists are more often men; Andree Collard, for example, counterposes masculine science to what she sees as women's values, asking 'Women especially must do some serious thinking and reconnect, if not to our gynocentric roots, at the very least to the history of man's violence to animals. For what has been done to animals has always preceded what has been done to us' (1988: 98). To her, scientists must be 'emotionally dead' if they can torture animals in the way that she describes (p. 67).

Becoming a scientist

Emotionality and empathy are qualities that, by contrast, a trainee scientist must learn to suppress; they get in the way of the quintessential 'objective pursuit of truth'. Somehow I knew, as a 17-year-old, that I had to swallow my disgust when confronted with a white rabbit with pink ears for dissection; after all, I wanted to be a scientist, didn't I? Learning to be objective means learning to distance yourself from those feelings. To become a scientist I had to leave emotion behind and learn to construct a façade of scientific authority. That is not to say that I no longer felt those emotions (though some scientists, perhaps, would deny that they felt them at all), but I had learned not to admit to them. I had also learned that, despite those emotional reactions, I did enjoy learning science. For all my disgust at seeing the rabbit, there was undoubtedly a fascination about observing first hand what the organs and tissues actually looked like, and how they fitted together.

For 'new recruits' to the biology laboratory, one important task is killing animals. In her analysis of animals as 'other' in feminist writing about science, Zuleyma Tang Halpin has suggested, for example, that an important rite of passage for trainee biologists is the 'pithing of the frog'. This task separates those 'who have what it takes from those who do not . . . The message . . . is clear: if you want to be a scientist you cannot let your emotions get in the way' (Halpin 1989: 286).

Objective detachment is, as feminist writers have pointed out, stereotypically masculine in our culture (Bleier 1984; Keller 1985), while 'not letting your emotions get in the way' is reminiscent of suppressing something feminine. Women in science have to take on board that fundamental contradiction, going through processes of desensitization towards animals is an essential part of attaining the required level of detachment. To identify with your animals (a more 'feminine' position) is to cease to be objective.

In making this point, I should emphasize that I am *not* saying that only women can identify with their subjects of study. What I am suggesting, rather, is that this feeling of empathy, of connectedness is gendered; it is stereotypically associated with femininity in our culture (and has been explored in explicitly feminist contexts). To become a scientist means acquiring some desensitization; it also means learning to speak and write in ways that gloss over some of the deep ambivalence that many scientists – male and female – often experience. This is how we learn in science to speak of 'sacrifices', or learn to make (unnecessarily) complex sentences in written papers that omit any reference to agency. Part of that learning process is to acquire the social skills of appearing not to be affected by emotions.

I can remember, for example, aged about 20, being in an undergraduate laboratory. We were asked to do experiments with frog skin, to find out about how different chemicals passed through the cells. The frogs (laboratory bred animals) had been pithed before pieces of skin were removed from their back; even though I knew intellectually that the animals were brain-dead, I felt sick at the sight of the pile of newly-pithed frogs. But I nonetheless swallowed my feelings of disgust, and I still conducted the experiment on my bit of skin. Meanwhile, I hid the emotions (or tried to; hiding feelings is not something I can claim to be good at). Social pressure to conform, to be a 'real scientist' is strong, perhaps especially so at that age.

Empathy versus experiment

Beliefs that empathizing with experimental animals is totally at odds with good experimental practice were evident very early in the Scientific Revolution. Some scientists worried that torturing animals might distort results; the microscopist Robert Hooke expressed doubt in 1665, for example, asking whether by 'dissecting and mangling creatures whils't there is life yet within them, we find her [nature] indeed at work, but put into such disorder by the violence offer'd' (Guerrini 1989: 401). Hooke may also have been more sensitive to the gendered basis of science than many of his contemporaries, including his mentor Robert Boyle. Boyle was both avowedly anti-feminist (Potter 1988) and quite prepared to justify animal suffering for the benefit of science (Guerrini 1989).

Scientific method was, moreover, itself portrayed at that time in terms of active male sexuality, 'thrusting' into her (nature's) secrets (Keller 1985). Most experimenters of the seventeenth century knew that the animals felt pain (despite Descartes's advocacy of animals as machines), but they believed that this was 'legitimately inflicted for the benefit of man and ... for the glorification of God through natural philosophy' (Guerrini 1989: 407) – nature yielding her painful secrets to the determinedly masculine methods of science.

Concern for experimental animals, on the other hand, connotes 'feminine' pity – an association that may well be one good reason for the repeated failures of anti-vivisectionist feeling. Women's involvement in the nineteenth-century campaigns led, perhaps predictably, to accusations of hysteria, or of being 'old maids' (Ryder 1989). More recently, the British entomologist Miriam Rothschild has described the process of desensitization that she went through in becoming a scientist, noting that other zoologists of her acquaintance feared being dubbed 'unmanly' if they showed compassion towards their animals (Rothschild 1986: 50).

This apparent gender contradiction did not, it seems, deter Rothschild, not least because students of biology were, she felt, brain-washed and 'disinclined to think'. It was another 30 years of research before she began to take the matter seriously . . . This fortunate, but traumatic experience I owe to my eldest daughter who, as a schoolgirl, resolutely marched out of her zoology classroom never to return. She refused to kill and then dissect an earthworm. The penny dropped.

(Rothschild 1986: 50–51)

Being (or trying to be) detached from the world that the scientist is observing is central to scientific objectivity; it is also stereotypically masculine. Gender connects to the issue of animal use by science in two ways; it is suspiciously unmanly to object to invasive procedures on animals, and ideas of masculinity are culturally associated with the objectivity demanded of scientists (Keller 1985). Women entering the life sciences have to overcome these associations. The problem besetting potential recruits to biological sciences is clear; it is you who has to change. If you object, or are too squeamish, then your only choice is to get out of doing biology altogether. At the moment at least, the practice of science is unlikely to change.

Empathy, a more 'feminine' characteristic, may not be explicitly acknowledged in scientific discourse; yet it is well known within the scientific community that some people are simply better than others at understanding or handling animals. These may not necessarily be women (although women working as animal technicians were felt by some scientists to be 'better' with animals in laboratory manuals earlier this century); an article published in 1947 suggested for example that 'the right technical assistant (preferably female)' was 'by far the most important feature in the management of any rat colony' (McGaughey *et al.* 1947: 111).

Similarly, Arnold Arluke (1992a) has described what he calls the 'ethical culture' of two very different primate labs in the USA. In one, the attitudes are cavalier towards the animals; the lab workers are 'cowboys'. In the other, the attitudes are more companionate and caring; these are 'animal people'. The gender connotations in his descriptions of caring animal people versus macho cowboys are clear. In her 'undercover story' for the British Union for the Abolition of Vivisection, Sarah Kite describes how laboratory workers forced tubes down beagles' throats for oral dosing of drugs, how terrified the animals looked, and how they were subsequently thrown back into their cages. The uncaring, 'macho' behaviour she describes seems to accord with Arluke's account, and contrasts with her use of the language of emotions – how 'terrified' the animals looked, for instance.

The image of the macho scientist is of someone who does not express feelings, may even seem cavalier in his (or her) attitudes towards the animals used. Researchers, of course, may sometimes seem cavalier out of ignorance; the philosopher and scientist Bernard Rollin suggests that they are often 'ignorant regarding many features of the animals they use, and receive little or no training in dealing with animals'; anecdotally, he cites a researcher who asked when mice grew up to be rats, and another who sent a batch of beagles back because they were feverish (dogs normally have a higher body temperature than we do) (Rollin 1989: 125–6). Such anecdotes abound, at least among those willing to acknowledge them.

I have no doubt that such scenes as the anti-vivisectionists describe can sometimes be observed: similar behaviour towards animals (and worse) can be witnessed in slaughterhouses, in intensive agriculture, in people's homes. Every day, the animal welfare organizations rescue thousands of ill-treated or unwanted animals – and destroy many. Perhaps ignorance is a factor, as is sometimes claimed; more likely, cruelty in, say, slaughterhouses, has to do with desensitization (which may be necessary to do the job) and the need of the people doing it to suppress their own ambivalence.

Despite the training that discourages the expression of concern, however, scientists are often as ambivalent in their beliefs about animals as anyone else. Outside of the laboratory, they may treat some animals as having moral claims (their own domestic pets, for instance). Yet inside the laboratory they may seem to behave as if the animals used in their experiments had few, or at any rate fewer, moral claims. Scientists using animals do, of course, largely accept that animals have less ethical value than humans; that is how research on animals must be justified in the first place. But less ethical value does not mean none, and many people working with laboratory animals do find ways of·expressing concern over ethics. While there is no doubt truth in the images of terrified beagles, and of uncaring people force-feeding them, many other people in laboratories seek to ensure the animals' welfare. Welfare, of course, is nowhere enough for those advocating animal rights – and caring for laboratory animals must be justified by those workers by reference to beliefs that using animals is permissible in the first place.

In their public pronouncements – in scientific journals, or in public defences of their research – scientists may seem to present a united front. The ethical issues emphasized in pro-research pronouncements from, say, the Research Defence Society, inevitably focus on the medical (and veterinary) benefits resulting from previous research. The discovery in the 1920s that insulin treatment could alleviate diabetes could not, they urge, have occurred without research on dogs, but has been of inestimable value to millions of human diabetics. Thus a publicity video produced by the Research Defence Society, 'What About People?', refers to the early research on diabetes and insulin, accompanied by footage of a diabetic man. He tells us how his life has been saved by regular insulin injections, as he pats his dog.

At other times, however, scientists may be more uncertain. They may express their ambivalence in speech, as conversation allows more scope for uncertainty or for referring to 'what can go wrong'. On one hand, scientists may use language clearly reminiscent of the 'air of bravado'. So what if an animal dies, 'there's more where that came from' I was once told, when an animal died on the operating table. That masculine bravado, not appearing to be affected by the death of the animal, is part of what the fledgling scientist must learn if she or he is to survive. Learning it meant that I had to learn to write in that 'objective', distancing style so characteristic of science; I had to learn to make claims about the relevance or importance (in scientific terms) of what I was doing. By contrast, when I worked with animals at home, training or living with them, I spoke a different language, a language rich with emotion and metaphor; even in the lab I would use that language to talk about the animals I worked with. One rat might be 'cheeky' for instance, another 'friendly'. But those descriptions did not find their way into the scientific papers. Only in writing do scientists routinely distance themselves.

Social relations in the lab: Scientists, animals and technicians

Scientists (myself included) have been trained in laboratories for so long that the places are taken for granted; what we were trained to do were experiments that would test hypotheses and (in time) generate 'facts'. Recent work in sociology of science, however, tends to tell a different story. Bruno Latour asks, for example, what it is that makes laboratories a source of political strength, helping to give science such power and authority. The answer, he suggests, lies in part in their reliance on 'inscription devices', the various machines and technology of scientific experiments (Latour 1983: 161). It does not matter what is being studied, all laboratories rely on devices that churn out written traces: graphs, tables of numbers, abstract symbols. It is these that create 'data', the facts of the scientific experiment, and help to create an air of authority about those data.

An important result of all these traces is that they position the observer (whether that is an observer of laboratory work, or a reader of the resulting written texts); no longer is the observer observing a scientist observing nature, but he or she must now watch numbers, and believe that these machine products represent raw nature. Once that is done, the observer is less able to contest the output; after all, the printed results are there for all to see.

The purchasing and use of these instruments shapes the social relations of the lab. Once bought, they must be used and their products (graphs, lists of numbers and so on) must be incorporated into a written product, the scientific paper. Producing this is one of the central acts of science (Latour and Woolgar 1979). For my purposes here, an important point is that they also inevitably shape the relationship between humans and non-humans, and thus the knowledge about animals that can be gained. They structure the space in which the work is carried out, and the work must be fitted into the demands of the technology. In interviews with animal technicians, for example, one told me that she had been expected to 'cull' all rats below a certain size (which included virtually all the females); only males over a certain size would fit the stereotaxic apparatus used to hold the head still during surgery for neuroscience experiments. The results, of course, are likely to be written up in such a way that they seem to apply to all rats, irrespective of sex.

Another example of how apparatus helps to construct (and limit) what we know about animals comes from studies of the sexual behaviour of rats. Until fairly recently, the literature was full of accounts of the passive receptivity of the female, who would be mounted actively by a male. One obvious reason why the active behaviour of female rats (who solicit males, choose between them, and help to 'pace' male activity; McClintock and Adler 1979) was ignored was prevailing ideology, the image of recumbent female sexuality. But another important reason was simply the apparatus; scientists had put the animals in heterosexual pairs in tiny enclosures to watch the sex. In larger enclosures, in groups as they would be in the wild, a very different picture emerges.

Physical space matters in another sense, too. It may seem a trivial point, but most 'laboratory animals' used for research do not generally live in laboratories. They live in specially-built animal houses, cared for by specialist animal technicians. Many animals never go near a lab; they are kept as breeding stock, fed and watered regularly. The design of animal houses and cages has come under increasing scrutiny in recent years in the wake of growing public concern and consequent legislative changes. Lab animals, then, live in cages in animal houses. What this means is that there is a spatial separation between the animal house, where the animals live and are cared for, and the laboratories in which many procedures are carried out. In part, this separation has evolved to ensure the health of the animals and their quality, but it also ensures that at least some laboratory personnel rarely, if ever, come into direct contact with the animals (or their caretakers) in their daily lives.

It is, of course, not only the physical space of the laboratory and its apparatuses that contribute to the construction of what we know about animals. It is also the people who work in them; at a general level, it is 'scientists' who are doing the experimental procedures on animals that the anti-vivisectionists abhor. They are scientists in that they are involved in a process called science. But so are the animal technicians, whom we do not usually call scientists. In practice, there are many people who are involved in 'doing science', from the lab director and the researchers at the lab bench to the technicians, as well as the secretaries and cleaners (not to mention the women whose domestic work reproduces the scientists themselves, as Hilary Rose has pointed out; 1983, 1994). Many of these people are not thought of as 'doing science', even if they are essential to it; others are doing it, but are not labelled as scientists (Latour 1987).

There may be others who are involved in the negotiations concerning what happens to lab animals. In Britain, for example, the use of animals in research is regulated by law, and managed through the Home Office and a system of inspectors. The law requires licences (for the institution, the researcher and the particular project); it also requires that there is a named vet associated with the institution. So the inspector is likely to be consulted as part of the process of obtaining licences.

Within this division of labour, there are different relationships to laboratory animals. Some scientists may have little to do with the animal house, the place where the animals are kept; instead, they might request that an animal is delivered to the lab for an experiment. Depending on the research, the animal may be killed first, or it may be sent to the lab alive and never allowed to wake up from the anaesthetic. Some may undergo surgical procedures (usually in an operating theatre) so that scientists can study the subsequent effects. For these scientists, the animal merely appears at a predestined time and place, as part of the apparatus. It has no meaning *as* a living animal, only as an exemplar of particular physiological systems or biochemical processes.

Some scientists, by contrast, do their work in the animal house itself. They may prefer to give animals injections rather than to ask a technician for example. Others work on topics that involve observations of the animals. My own background, for instance, is in animal behaviour. I conducted research by observing animals in the animal house where they lived. The meaning of the animal for those scientists who do so is likely to be different from the meaning of the animal if it simply appears in the lab. Watching animals daily going about their behavioural routines meant that I could not easily see them as bits of apparatus. Like many technicians who work with animals, I would often go into the rat room in the morning and greet them all with a cheery hello. You don't say 'hi' to your ultracentrifuge.

Animal technicians must care for stock animals, deliver animals to the labs, and care for animals recovering from surgery. They must set up breeding programmes, and may also take part in some of the research (if their involvement is permissible under the laws or guidelines of the country in which they work). They must also cull animals that are too old, or if there is overstocking.

The technician is thus closer to the animals than most scientists; not surprisingly, many feel therefore that they are 'buffers' between the animal and the scientist. They see themselves as caring people, who come into the job because they like animals. Because of that, they may also feel that they are the 'best ones for the job' when it comes to the unpleasant task of having to kill their charges; no one else, after all, knows the animals as well as they. One technician told me, for example, that the ferrets she worked with 'were her friends'; she knew them best and they trusted her. Paradoxically, this meant that she was the best person to deal with them 'when the time came'.

Technicians must care for the animals as individuals. They may empathize with the animals, yet must justify the research. But they can do little about these contradictions (even if they find ways of quietly resisting); they lack the power that the research scientists have. Similar questions of power are familiar in feminist writing; what structures the division of labour in the laboratories is, inevitably, gender, class and race. Technicians are more likely than research scientists to be women or to be black, for example. The relationship of each of these actors to the laboratory animals may be different; but each relationship is situated within a wider public arena, and in different relationships of power. Scientists (and inspectors) have more power than technicians and are more likely to be male. Thus relationships of humans to animals acquire an overlay of gender – though the end result for the animals is, of course, always the same. Laboratory animals are at the bottom of the social pecking order.

Most of those involved in research, not surprisingly, tend publicly to defend it, but few privately are that sanguine. In private, scientists sometimes also express feelings about the animals, and acknowledge the need for empathy and 'good handling' (Lynch 1988), even though open discussion of these feelings is not encouraged. Talking publicly about scientists' unease can meet resistance in scientific circles; Arnold Arluke encountered suggestions, for example, that he change the title of his talk to one that was less 'provocative' than one expressing unease. In the end, he called it 'untitled' (Arluke 1992b).

Similarly, scientists in conversation may seek to convey an impression of themselves as caring, emphasizing for example ways in which they draw the line at particular techniques, at using particular species, or at the use of animals for testing cosmetics for instance (Birke and Michael 1992a, 1992b). Others may state that while they could justify using rats or mice, they could not bring themselves to work on dogs, cats or primates (noting as they did so, that they were being hypocritical or contradictory in adopting this position). Using euphemisms, too, is a way of avoiding admitting to the dilemmas caused by using animals. While some animal technicians do use the word 'kill', they are far more likely to use 'cull' or 'sacrifice' or 'put down' – even when I had explicitly used the more direct word in posing the question.

Perhaps one reason why scientists often find it easier to work with rodents rather than larger animals is that rodents are relatively stand-ardized. At least to the untrained eye, one white rat looks much like another. They therefore lose individuality (unless singled out in some way for special treatment, when they may cross the boundary and may even become 'pets'; Arluke 1988). I well remember one rat in our lab who had problems with his teeth. Rodent incisors keep growing, but his were slightly skewed, so they failed to meet properly and were in danger of circling round and through his jaw. We didn't 'cull' him; instead, we regularly trimmed his teeth so that he could eat. Somehow, because of his infirmity, we singled him out, and created a pet; we were upset when he died. Yet we gave none of his sisters and brothers any special treatment. They remained numbers in cages, part of the experimental protocol, each much the same as another. We were not unique in this strange behaviour of creating 'pets'.

In one visit to an animal house, I was struck by the way that the animal technicians spoke of the need for opaque cages. The reason, they said, was that the research scientists felt 'disturbed' by seeing the rats in clear cages, 'because they kept looking at you'. Once they do so, animals begin to cross the boundary from being 'apparatus'. In this sense, they become less clearly 'other' when they start to watch you – they become individuals, real animals. For these laboratory scientists at least, there are clearly ethical dilemmas – better not to see that they are animals at all.

The relationship of scientists to the animals under investigation is somewhat different when the animals are being studied in the field. In the first place, the animal populations are usually being studied for their own sake, as exemplars of a particular species or as part of a particular ecosystem. Not surprisingly then, field studies cover a much wider range of species, from nematode worms to chimpanzees. Laboratory studies, by contrast, are typically based on cheap and easily bred small mammals such as rodents. In many laboratory based studies, the animals used are frequently standing as a substitute for humans – the 'animal model' of mammalian physiology. They are, furthermore, artefacts of laboratory practice, having only a symbolic relationship to 'nature' outside of the laboratory. Animals in the wild, unlike laboratory animals, clearly have a less symbolic relationship to wild nature; they are also more likely to be seen by scientists as whole animals, rather than as suppliers of bits. Second, the ethical dilemmas are different. A scientist (or others) might be concerned about inflicting possible pain or suffering on the animal(s) concerned, in much the same way that someone might be concerned about laboratory animals. This would be the case, for example, if a field study involved capturing an animal and subjecting it to an injection, say, or to minor surgery. There are other ethical dilemmas involved in field studies, however. Should a scientist be allowed to remove one animal from a population that is breeding? What happens to that animal if moved elsewhere? What happens to other animals in the population, including that individual's mate and offspring? And, of course, these problems may be compounded if the species in question is endangered.

A third difference between field work and what goes in the lab is one of degree rather than kind. The relationship of scientist to animal is, as I have said, embedded in a wider social context of relationships between people which both structure it and give it meaning. Unlike lab studies, which tend to be done most frequently in affluent western countries, some of the best known field studies take place in poorer, developing countries. In that case, the researcher enters another culture. In primate studies, for example, Donna Haraway has pointed to the ways in which the research lies at the intersection of politics of gender and race. What we know, scientifically, about chimpanzees or gorillas has been obtained predominantly by white people (significantly, often white women) whose presence in the African countries where the animals live depends upon a long history of western colonization of those countries. How we think about other apes, in their taxonomic closeness to us, is also part of the politics and negotiation of the meanings given in western science to 'nature' (Haraway 1990).

Defending the boundaries: The animal experimentation controversy

It is, of course, precisely their closeness to us that makes the apes special in our eyes. For that reason, many people are more opposed to using them in laboratory work of any kind than other species of mammals. Not surprisingly, the use of primates has been an important theme in the growing controversy about the use of animals in laboratory research.

Since the mid-1970s, there has been such growth in anti-vivisectionist feeling that we can speak of a new social movement for animal rights (Jasper and Nelkin 1992). Scientists can no longer dismiss the opposition as merely a lunatic fringe; they ignore anti-vivisectionist feeling at their peril. Increasingly, they must act to protect their laboratories and animal houses (with the high protection described so caustically by animal rights activists), and they

> must engage in a battle for public opinion, seeking to undermine the animal rights movement by associating it with violence and terrorism, and to improve their own public image by defending the value of their work. And they try to steal back the moral high

ground with emotional rhetoric. In effect the scientists have developed a counter-crusade.

(Nelkin and Jasper 1992: 38)

Thus in 1990 more than one thousand people signed a declaration on scientists' use of animals organized by the British Association for the Advancement of Science. The subsequent short report (published in 1993) stressed the benefits to medicine of animal research, and included a special section on the importance of using primates for certain areas of research such as vaccines (AIDS is mentioned here) and Alzheimer's disease.

A key strategy of this report is to emphasize that scientists have clear responsibilities in using animals. 'If there were a Hippocratic oath for scientists', the report argues,

> it would surely include two primary responsibilities that a research worker must accept when planning animal-based studies. The first . . . is to use animals only for research intended to contribute to the advancement of knowledge. The second is to minimize any possible pain or distress that an animal in the scientist's care may experience. Those are two of the key criteria taken into account when deciding the scientific merit of a research project and before a request is made for funds for the research.

(BAAS 1993: para. 5)

My main point here is to emphasize this text as part of a counter-strategy on the part of scientists. Note the stress placed in this writing on responsibility and on using animals only for certain purposes and under certain conditions. This rhetoric is designed to reassure waverers in the controversy that animal experiments are not done for trivial reasons (as is often claimed by animal rights protesters). It will do little of course to mollify the protests. After all, it is quite easy to claim that a research programme is *intended* to contribute to the advancement of knowledge; what self-respecting scientist would claim otherwise, when advancing knowledge is the name of the science game? And minimizing distress is in the scientists' interests as well as the animals', because it helps to keep unwanted variability in the results to a minimum.

The counter-strategy draws us back to the kinds of critiques that both feminists and animal rights protesters have made against science. Thus, while individual scientists may indeed be responsible, collective responsibility is less likely in a system that so subordinates the animals' interests. Appeals to the responsibilities of scientists focus on the individual, removing him or her from the social relationships of the laboratory, and from the economic and political structures in which research is done. In practice, it is not likely to be the research scientists themselves who will oversee the day-to-day welfare of the animal (even though they may be legally obliged to take that responsibility). And no funding agency is likely to give money to a research project that did not appear to be 'advancing knowledge'.

A less obvious counter-strategy is to defend one's own research in contrast to that done by others. Here, the trick is to admit that 'abuses' of animals in research might occur elsewhere while believing that everything 'here' is all right. The 'here' might be one's own lab, or country, or social group; it is always 'others' who perpetrate misdeeds. Given the high profile of the controversy, it is hardly surprising that researchers would be loath to recognize problems in their own lab, while admitting that they may occur elsewhere.

Here, the 'other' distinction is not between humans and other animals, but between the scientist him/herself and other people who might, in some contexts, be seen as part of the same social group. These others are ones whose treatment of the animals is held by the speaker to be insufficient or cruel. Thus others might include 'foreigners' (explicitly racist or ethnocentric), that is, scientists from those countries or cultures whose attitudes to animals may not be the same as the speakers (Birke and Michael 1992b). It might be petkeepers among the lay public, who often fail to live up to standards they expect of scientists. These 'others' serve the purpose of marking out ethical boundaries; for the British scientists we interviewed, it was British science that was seen to be better – more ethical, better controlled – than science done elsewhere.

One way of surreptitiously policing the boundaries so that anyone with suspect tendencies towards feminine empathy for animals is excluded is simply to expect everyone within science to adopt an air of bravado, of not asking awkward questions – or not in public, at any rate. Meanwhile, the 'sentimentality' (or caring empathy, if you will) of people outside of science is scorned in pronouncements by the research establishment, and people are derogated as being 'ignorant' of the benefits of animal-based medical research. Ironically, that establishment is beginning to fight back not on grounds of scientific reason, but of emotionality itself. People in wheelchairs or children saved by the products of research provide emotive images and language that can counter the gruesome pictures of anti-vivisectionist literature. As the American Medical Association put it in 1989, scientists must fight back using tactics as hard-hitting as their rivals, for 'the general public is up for grabs' (Jasper and Nelkin 1992: 132).

It is, of course, precisely appeals to emotions and 'feminine sentimentality' that have had currency in feminist writing about the use of animals. Sally Gearhart, writing in the newsletter of Feminists for Animal Rights (Gearhart 1992: 1–4), points out that scientists may not be intentionally cruel, but they have become dehumanized. That dehumanization, that lack of caring compassion, promotes a deep alienation of scientists from the life of the nature they purport to study; that alienation concerns Gearhart, because of its potential links to wider domination. Alienation breeds violence, she implies.

My purpose in this chapter was to explore two themes about the laboratory use of animals that raise feminist questions. One is the social relationships of the lab, not only between people in different occupational positions, but also between people and animals. The meanings of the animals (and the practice of what happens to them) are structured by those relationships, which in turn are deeply gendered.

The other kind of feminist question focuses on the gendered nature of scientific inquiry and practice. Caring empathy is, as Miriam Rothschild noted, 'suspiciously unmanly' and weak, while the façade of emotional detachment is one that we must learn if we are to survive in science. Feminists have questioned those stances, and the underlying assumption that detachment is more valuable as part of our broad critiques of science. Scientific writing, meanwhile, continues to perpetuate a language devoid of emotional nuance or richness of metaphor – a language, indeed, that is largely devoid of any sense that there are sentient animals in the laboratory at all.

5

WRITING THE ANIMAL

... human beings sometimes display an incredible viscosity of imagination. When all you have is a hammer, everything sooner or
later begins to look invitingly like a nail. Some of the simplicity
physicists find in nature may be the result of the procrustean tools
(currently the language of quantum field theory) they use to work
with nature.

(Gregory 1990: 166)

Physics, to many people the epitome of hard science, is 'only indirectly
about the world of nature', Gregory contends. 'Directly, it is talk about
experimental arrangements and observations ... What is not given to
physicists by nature, but rather is invented by them, is what they *say*
about these outcomes, the language they use to talk about nature' (1990:
181). What scientists say – in print or in conversation – reveals much
about what the natural sciences are. If we are to believe the popular
conception of science, these words describe a reality, a truth; they directly represent nature. But do they?

As science itself has come under increasing scrutiny from sociology
and linguistics, other accounts emerge of what scientists are doing.
Scientific research papers become not straightforward accounts of doing
an experiment, but an exercise in rhetoric (Gross 1990; Latour 1987).
Not only that, but it is, as Latour points out, a very peculiar one; it is
constructed in such a way that there will be few readers who stay the
course and read it right through. Its very style seems to chase readers
away. 'Made for attack and defence' against potentially hostile readers,
Latour suggests, 'it is no more a place for a leisurely stay than a bastion
or a bunker. That makes it quite different from the reading of the Bible,
Stendhal or the poems of T.S. Eliot.' (1987: 187).

Bearing this in mind, I want in this chapter to look at the language of scientists as they write or speak about experiments using animals. The practice of biology, no less than that of physics, is 'only indirectly about the world of nature'. The accounts biologists give in their written papers are carefully contrived, constructed after the event, using language designed to parry potential criticism. What happens to the animals in these accounts? Does the possibility of criticism about the use of animals affect the rhetoric? And what does the animal in the laboratory represent?

Who, or what, is the laboratory animal?

Science purports to study nature. Even in the laboratory, the animals, plants, tissues, cells or whatever else are studied, are all taken to represent similar structures or processes in nature. But how much does 'nature' enter into the world of the laboratory? How representative of animals 'in nature' are the ones that are used in laboratories? Indeed, who or what *is* the laboratory animal, and how do scientists talk about it?

The combination of public opinion and cost ensures that relatively few of the animals used in research worldwide are creatures such as primates, dogs and cats. Most (as defenders of research often point out, playing on public ambivalence about 'nastier' animals) are rats and mice. These are, after all, easier to convert in the imagination into the 'analytic animal' than, say, a dog. But these rodents are not ordinary rats or mice; these are animals especially bred (indeed, designed) for the purpose.

'The laboratory animal' becomes a tool in the service of science, part of the apparatus. Indeed, advertisements for various specialized strains of laboratory rats and mice make this clear; 'purpose bred . . . for all your experimental needs' is a theme of such adverts, which may pictorially juxtapose images of the rodent next to one of the working scientists or his/her apparatus. Breeding programmes have created uniformity, to remake animals as though they were simply another laboratory reagent (Lane-Petter 1953).

Laboratory animals are both literally and symbolically created; most are constructed in the sense of being specifically bred *as* laboratory animals. Many laboratory strains of purpose-bred animals were developed in the late nineteenth to early twentieth centuries. Breeding stock of several species derived at that time from animals bred by lay 'fanciers' (Worden 1947) who also contributed much knowledge of breeding and inheritance. Science however required uniformity – the 'fancy' varieties of rodents were 'less satisfactory' for lab use (Wright 1922) – so efforts were made to create standardized breeds. In that sense, laboratory animals were literally constructed to fulfil a specific role in science.

This underlines a point that sociologists of science have emphasized, that the laboratory is a highly preconstructed artifactual reality . . . the source materials with which scientists work are specially grown and selectively bred . . . 'Raw' materials which enter the laboratory are carefully selected and 'prepared' before they are subjected to 'scientific' tests. In short, nowhere in the laboratory do we find the 'nature' or 'reality'

that science purports to study (Knorr-Cetina 1983: 119). Raw nature is largely excluded from laboratories.

It is not part of the animal house either; the animals are purpose-bred and environmental conditions in the animal house carefully controlled and monitored. In turn, such stable conditions can help to ensure the health of the animals. Physical health, at least, is vital for science. Not surprisingly, then, manuals give fine details about the physical environment of the animals: cage sizes, air quality, feed, bedding and so on. If these are well controlled, the environment is less variable; if variability is reduced, so should the physiological variability of the animals be reduced.

The controlled conditions and selective breeding ensure 'quality products' that can seem to bear little resemblance to the animals we might meet elsewhere. 'Nature' does indeed seem to be far removed from this scene. At an extreme are the units producing 'pathogen-free' animals – animals raised and maintained in completely sterile conditions. A more unnatural animal can hardly be imagined.

Ironically, although breeding programmes have sought to develop the 'uniform' laboratory animal, we know much less about the living conditions that might produce stress in the laboratory itself. Some things seem obvious. If a dog, say, were kept in too small or cramped a cage then it might well develop behavioural problems. Some are less obvious, such as the effect of ultrasound on rodents. Gill Sales and her colleagues at the University of London have looked at the effects on rats of being exposed to ultrasound, the kind that is routinely emitted from a computer monitor for instance. Male rats became much less active in the presence of monitors, but females did not, she found; moreover, female rats exposed to pulses of ultrasound did much less rearing up and looking around (Sales 1988). Computers are standard equipment in laboratories and offices; what then are they doing to the rats living nearby? And if they are stressed, what are the effects on their physiological systems?

There is thus a gap between the tightly controlled conditions that scientists claim and what is produced in the form of scientific knowledge. A big problem in doing biology is the variability inherent in the raw material, which cannot be controlled. This inevitably shapes what emerges in the form of results. Karin Knorr-Cetina cites a study of neuroscientists using rats; the problem with rats, however carefully bred and housed, is that they 'tend to squirm, squeak, kick, try to bite, and wriggle free.

Thus they tend to cause troubles which require unmethodical remedies', she points out (1983: 124). These remedies may well not appear in the final report, which will be a model of the abstract writing and apparent certainty that we might associate with scientific writing.

The behaving animal can thus mess up the apparently tightly controlled experimental procedures. What animals do in their cages can also disrupt results, particularly if bored animals develop stereotypies or other abnormal behaviour (I make no apologies for describing them as bored; it goes against the spirit of this book to assume – as my scientific training would have me do – that boredom is a term that should not be applied to non-humans).

The behavioural needs of laboratory animals are becoming increasingly recognized, not least because an animal that develops stereotyped patterns in its cage is likely to be a stressed animal. And if it is stressed, then this in turn could invalidate results (Fox 1986). One way in which those needs might be met is that housing conditions can be improved; simple changes include making cage sizes larger, or allowing animals to live in groups as their wild cousins would do. Environmental enrichment is another possibility; having to search for food items in sawdust, for instance, may more closely approximate what a monkey would do in the wild. In one study of the effect of such enrichment, the researchers found that macaque monkeys spent 5 per cent of their time searching through woodchips on the floor, even in the absence of food. If food grains were added, they spent 15 per cent of their time searching; and if no other food was available, this rose to 30 per cent of their time (Chamove *et al.* 1982).

My concern here is not to review the literature on this theme, but to make two points. First, these changes make subtle differences to the ways in which the people working in laboratories see the animals. Animals viewed merely as numbers in small cages are more easily seen as laboratory apparatus than an animal with whom you interact in the process of providing environmental enrichment.

Second, such changes are, symbolically, at least, a means of bringing 'nature' into the lab, if only a little. Environmental enrichment seeks to encourage the 'natural' behaviour of the animal, to discourage the development of abnormal patterns of behaviour. The animal itself may be far removed from wild nature, but it can be encouraged to behave as though it is not.

Researchers can make themselves feel more comfortable with the dilemmas of using animals if they can symbolically (as well as actually) move 'the animal' from the realm of common sense to the realm of data. The trouble is that animals (particularly mammals) have a nasty habit of reminding you that they really are animals after all. Looking at you through the clear plastic of the cage is one way; behaving in ways that somehow single the individual out from its peers is another. It is much harder to think of them as analogous to laboratory apparatus when they do that. Moreover the more that researchers working with animals provide environmental enrichment, the harder it is for them to escape the observation that these are, after all, not apparatus but animals, whose behavioural and physiological needs may, or may not, be met.

That ambivalence among researchers inevitably influences how they talk about animals. Conversationally in the lab they may well recognize the sentiency of the animals they work with, but when it comes to written papers, such recognition drops away entirely. The written paper is a model of denial.

The language of science

> Experiments were carried out on 107 healthy Sopranoes (*Cantatrix sopranica* L.) . . . weighing 94–124 kg . . . Halothane anaesthesia was utilized during the course of tracheotomy, fixation . . . and major operative procedures. 5% procaine was injected into skin margins and pressure points. Animals were then immobilized with gallamine triethyiodide (40 mg/kg/hr) and normocapnia was maintained by appropriate artificial ventilation. Spinal cord transections were performed at L3/T2 levels, thus eliminating blood pressure variations and adrenaline secretion induced by tomato throwing . . . The fact that the animals were not suffering from pain was shown by their constant smiling throughout the experiment. Internal temperature was maintained at 38°C ± 4°F by means of three electrically driven boiling kettles.
>
> (Perec 1991)

In this spoof of a scientific report, Georges Perec parodies the language and style of writing that typify science. Sopranoes, it suggests, show a yelling reaction if tomatoes (or other missiles) are thrown at them, and various scientists have sought the neurological basis for this reaction in different parts of the Soprano brain. Although clearly a spoof, Perec adheres closely to the kinds of statements, language and structure of arguments typical of neuroscience. The paragraph quoted above, from the 'Methods' section conveys with a black humour the ways in which the 'preparation' of experimental animals is typically described.

In an analysis of the 'shop talk' of scientists working with animals, Lynch (1988) distinguished two different ideas of 'an animal' that were evident in scientists' conversation. The first was that of a 'naturalistic animal' – the animal of common sense to which most of us might refer when we talk about animals conversationally. This is the animal as it might be found 'in nature', or even as a pet. It can be found in the practice of science – technicians may use this sense of 'animal' in their work, for example – even if it rarely finds its way into journal articles.

The naturalistic animal, we presume, is probably sentient, and is at least capable of pain and suffering (or at least we assume that for most mammals and for some birds; popular perceptions of other kinds of animals are more uncertain. I am not clear, for example, whether some venomous snakes would fall within the category of 'naturalistic animal' in the way that Lynch associates it with common, or common sense, usage).

The second sense of 'animal' used in science is what Lynch terms the 'analytic animal'. This is the animal transformed into data (or into artifacts such as electronmicrographs); he notes how scientists looking at photographs of cells taken from the brains of rats would say such things as 'That was a good animal, that was'. What they see is not, of course, an animal or even a direct representation of one. It is a part of an animal that has been *transformed* into data. Whatever else a laboratory animal might be, what it becomes is data. How scientists talk about science is an indicator of how they perceive the dilemmas; so, too, are the written accounts. The unease of laboratory workers does find its way into scientific writings – the use of the word 'sacrifice', for example, to describe killing is commonplace in the published accounts (Arluke 1988; Lynch 1988; interestingly, my impression of reading a few laboratory reports written in the earlier part of this century is that the word 'kill' was used more readily; this might reflect an assumption that the readership was likely only to be other scientists, or it might reflect the fact that, during the middle years of the century, anti-vivisectionist feeling was at a relatively low ebb).

Yet on the whole, scientific writing conveys an image that is much less ambivalent than conversations may be. In written accounts of science, feelings have no place, and their language and construction help to create this distancing through the use of particular words, or the use of the passive voice, for example. These help to create the 'missing agent'; there is no person, merely a procedure or perhaps 'an inability to assay' (in this case, the animal died in vain; it cannot become data if someone messed up the subsequent assay). In scientific reports, 'lesions' or 'sacrifices' seem just to happen to animals, without any human agency at all (Lynch 1988).

Not only is it freed of human actions; omitting agency also glosses over the messiness of day-to-day science. Lynch observed scientists through a series of experimental procedures, noting various ways in which the procedures were varied in response to changing conditions: the perfusion did not 'take'; the rat did not go under the anaesthetic properly; the needle jammed. Yet, he notes, the research reports was written up to describe animals 'sacrificed under Nembutal anesthesia by intracardial perfusion utilising a mixed aldehyde fixative media' (Lynch 1988: 74; what this means in English is that the animals were killed by a technique called perfusion. First, the animal is anaesthetized using a drug called Nembutal; then, while the heart is still beating, the scientist injects directly into the heart a 'fixative' containing chemicals called aldehydes. This will kill the animal, but it also serves to 'fix' the body tissues – in this case, the brain – as it is pumped around by the heart. This prevents the shrinkage of brain tissue that might result if the animal was first overdosed with anaesthetic and then its brain removed for analysis). The image that is created in conversations is one of a kind of machismo brutality; 'this one's fucked', says one researcher to the other, while he repeatedly stabs the rat with the injection needle.

The writing, moreover, is itself constructed in ways that diminish the significance of the animal. Scientists are trained to write in the passive voice, which removes the practice of science – and the scientist – from the sentence, so reducing the emotional impact on the reader of what is done to the animal. Scientists do all kinds of things with the animals they use, and with their tissues, but, in the written account, human agency disappears. It is noteworthy, however, that in an analysis of the content of scientific papers Bazerman (1988) noted that the uses of active verbs occurred almost entirely in connection with intellectual processes. The scientist may actively think and generate hypotheses, meanwhile, the animals are miraculously anaesthetized or injected by an invisible hand.

Erasing the human agent by the consistent use of the passive voice moreover helps to perpetuate a gendered account of science. Like the generic 'he', a sentence having no pronoun at all is likely to be read as implying male actors, as will a sentence in which an active verb implies intellectual processes. If animals were injected by invisible hands while the scientist thinks, then we might wonder whose hands were doing the injecting.

Another diminishing tactic is to talk in phrases such as 'the guinea pig gut'. In this case, a small fragment of an animal is isolated, and measurements taken; what constitutes data are the products of the various machines that are linked up to the animal fragment (Latour 1987: 66). The scientific paper must be written in such a way that a central role is given to these machine artifacts – the data. The living animal, whose 'sacrifice' provided the fragment of gut, is unimportant.

Scientific papers are, in a sense, fictionalized, idealized accounts in which 'style . . . is not a window on reality, but the vehicle of an ideology that systematically misdescribes experimental and observational events' (Gross 1990: 84). Lynch has similarly contrasted the idealized account of the written records with what he terms 'laboratory shop talk'. Conversation, unlike written records, typically makes reference to 'what can go wrong'. It can also make reference to a more empathic feeling for the animals. Wieder has described, for example, how scientists working with chimpanzees may make reference to the cognitive abilities of the animals in conversation, but this awareness is lost in written accounts (Wieder 1980).

Many journals do, of course, require that authors explain how many animals they used, of what species, and so on, in some detail. Some now insist that submitted papers adhere to published guidelines regarding how animals are kept or used. These procedures do help to make the animal more present. But few papers give full details of, for example, the housing and husbandry conditions of the animals even though these might be significant variables in experimental design, influencing the animal's physiology and thus the experimental outcome. Moreover, producing papers according to strict formats – introduction, methods, results, conclusion – may serve further to reduce the significance of the animals themselves.

Bazerman (1988: 260) points out, for instance, how increasingly prescriptive requirements for publication in the *American Psychological Association* journals may lessen

the likelihood that researchers will consciously consider the exact significance of . . . information, whether it and other possible information should be included, and exactly how this information should be placed in the structure of the whole article.

In other areas of biological research, where the animal is even less of a 'subject', the impact of such unconscious choice of linguistic structure may be considerable. Thus, the centrality of 'the gut' rather than the guinea pig is reinforced by the way the paper is written; it is the readout from machines that will dominate the methods section, not the animal who lost its life.

'The animal that lost its life' connotes a furry creature, a guinea pig, similar to the kinds of animals some people keep as pets. Not surprisingly, this kind of image is not likely to be invoked in scientific reports. It is the image of the 'naturalistic animal' of common sense (Lynch 1988); it is also the animal whose death is more likely to disturb us. So, scientists will sometimes admit to being troubled by the death of a particular animal during an experiment. For me, it was the death of a guinea pig because of anaesthetic problems that upset me particularly; I saw it as an individual, a furry creature, whose death (however inadvertent) had been caused by my actions. Yet scientists must, at the same time, accept the routine deaths of the many animals that are 'culled' in the background by technicians as part of stock maintenance; those deaths are the necessary price of having animals in the lab at all.

These animals are surplus stock that must be killed, or discarded. The language is significant; consider the impact on a non-scientific reader of the language used in the following extract, describing how the scientists examine the animals' brains after an injection was given directly into the brain: 'After sacrifice, the injections were verified; animals not showing the trace of the needle inside the ventricle were discarded (5–10%)'.[1] Here we read of 'animals' being 'sacrificed'. We read, too, that some were subsequently 'discarded', not used as data (helpfully, a percentage appears; as the number of animals used in the experiment was not given in this paper, however, it is not clear what the percentage represents).

Animals that are 'discarded' include, then, those who do not provide the kind of data that are required. In the quotation given above, these are animals whose brains do not provide empirical evidence that the injection needle had entered the required part of the brain; the subtext here is that this group will include those brains in which the missing agent – the scientist – failed to direct the needle in the appropriate plane. These are, moreover, no longer 'animals'; what is discarded here is not the whole animal (whose remains have long since been thrown away or macerated) but slices of brains. What the sentence refers to, then, is the discarding of particular slices from particular brains that did not show the needle trace (Lynch 1985).

Discarded animals also typically include those who did not die because of the scientists' action. Indeed, Lynch argues, it is only through such 'sacrifice' that animals *can* become data. The animal that merely upped and died anyway cannot do so; it does not enter the culturally symbolic process of ritual sacrifice (Lynch 1988). Not that these symbolic processes occur only in science; animals whose bodies are destined to become meat must similarly not die of their own accord. Fiddes (1992) notes how strong is the belief that, to become meat, an animal must be slaughtered; thus, meat-eaters are often loath to consume the flesh of animals that have died of other causes, even when those causes are known.

One characteristic of the animal-as-data is that it is less variable, less messy, than real, naturalistic animals. The use of tables, figures and so forth in scientific reports might even be said to *impose* invariance on the picture of nature that is being constructed. These form a pictorial representation that is

a triumph of simplification . . . not mice, but their brain cells . . . In tables and figures most of the properties of the actual physical objects, of mice or men, have been discarded, and all that remain have been normalised, ideally through quantification.

(Gross 1990: 74–5)

Scientific writing has become increasingly codified, particularly since the nineteenth century (Bazerman 1988). Although it is not explicitly stated, it seems reasonable to suppose that this formalization, through which the presence of the real, live animal is obscured in the writing, has been in part a response to criticism of the use of animals. This may be partly unconscious, or authors may play down what they did to animals because of their own ambivalence, or their fears of reprisals. It may partly be the result of pressure from journal editors. Susan Lederer (1992) has noted how, even in the first part of this century, when anti-vivisectionist activity was less obvious, journal editors rephrased submitted papers so as to reduce the emotional impact of what had been done to the animals.

Written descriptions of field studies present a somewhat different picture. The passive voice still prevails, but the animals tend to have more agency; this is a world in which chimpanzees, gazelles, geckos, or termites all do things. They are, as Gross emphasized, subjects of a sentence but they are (sometimes) subjects of a life too. They may even be granted individuality; primatologists working for long periods in the field often refer to individual animals by name – Dian Fossey's gorillas and Jane Goodall's chimpanzees appear in books very much as individuals.

Yet there is common ground. Reports of field studies are increasingly located in an abstract realm of mathematical modelling, attempts to describe the dynamics of animal societies through the medium of mathematics. Despite the description of the idiosyncrasies of some individual animals (usually primates), the focus of most field studies is on groups of animals. Like the animal in the laboratory, 'the animal' of field studies is usually an exemplar of a particular species; it is not individuals, but species or populations, that are theorized. The combination of a focus on species/groups, and of the abstractions of (say) mathematical models in evolutionary theory serves to distract attention from other sources of variance (individual idiosyncrasies included) – much like the effect of creating invariance that Gross suggested for the use of tables in laboratory reports.

The relationship of scientists to 'the animal' in field studies is as much about 'the other' as it is in the laboratory. The scientist must still stand outside, dispassionately watching the animals (even if that is much less possible if you are studying primates). The narratives, of course, differ:[2] 'animals' in field studies represent 'the wild' – wild nature as other to (largely western) culture. Animals in the lab are nearer to laboratory artefacts than they are to whatever we mean by wild nature.

But are these feminist questions?

In these chapters, I have alluded to bodies of literature that have dealt with how science is written and talked about by scientists, and how this contributes to particular beliefs and notions about 'animals'. This literature makes little, if any, overt contact with feminist critiques. But there are, it seems to me, several important questions for feminist thinking about feminism and science. Here, after all, is the heartland of scientific naming and defining; the animal is defined as determined by its genes or hormones and from there it is a small step to extrapolate to humans. That process of extrapolation is the basis of the biological determinism that feminists have so often had to criticize.

A clear theme in this literature is the contrast between what seems to be machismo, bravado and a more empathic quality. The macho scientist does not express feelings, may even seem cavalier in his (or her) attitudes towards the animals used. Yet both attitudes can coexist in science; some labs may have what Arluke (1992a) calls an ethical culture, staffed by 'animal people'. Others may be more cavalier, uncaring – the 'cowboys'.

How scientific narratives construct our perceptions of laboratory animals and what happens to them is also a feminist question, I would argue. The practice of science is deeply ambivalent. On the one hand, scientists must assume similarity between humans and laboratory animals if they are to justify their research to many of the grant-giving bodies (let alone to the public at large). On the other, it is difference that underlies the decision to use other species. This tension – and the ethical dilemmas that are founded upon it – is played out in the ways that scientists operate.

Consider, for example, the social division of labour: research scientists do the experiments, while animal technicians care for the animal after surgery. The responsibility in British law may lie with the scientist who holds the government licence. In practice, at least some of that responsibility for the ongoing welfare of the animal lies with the technicians. The scientist thus avoids any emotion-jarring empathy, any sense that the rat might just be distant kin. Instead, researchers can operate by assuming ethical, if not biological, difference; they can thus assume indifference (note that I am not suggesting that scientists are necessarily indifferent to the welfare or fate of their animals; at the very least, the animals' welfare is essential to the outcome of often costly research. The indifference I invoke here is towards the possibility of similarity between animals and humans, which can in turn provoke emotional identification with the animals).

Technicians, on the other hand, must care for the animals as individuals. For them, both similarity *and* difference have salience. They may empathize with the animals, yet need to justify the research by appeal to difference. But they can do little about these contradictions (even if they find ways of quietly resisting); they lack the power that the research scientists have. Similar questions of power are familiar in feminist writing; what structures the division of labour in the laboratories is

inevitably gender, class and race. Technicians are more likely than research scientists to be women or to be black.

The narratives of science, as we have seen, serve to exonerate scientists
by simply erasing or downplaying what has happened to the animals.
The animal, as a living being, is discursively restructured to become part
of the apparatus of science – the supplier of the gut, or of the neurons.
If animal experiments are the concern of feminists, as several writers
have argued, then we should be looking critically not only at what
scientists are allegedly doing to animals – implanting electrodes, depriving
them of sight, poisoning them with toxic chemicals, or whatever. We
should also be analysing the contexts in which these practices may (or
may not) be occurring, and how scientists account for them.

Where animals are located in science – both literally, in laboratories,
and figuratively, as the subject matter of science – is an important part
of the way that science itself helps to perpetuate the notion of human
superiority over other creatures. These creatures are often sentient,
intelligent beings, even if very different from us (I am cautious here; I
do not know how many animal kinds we could think of as sentient.
Perhaps a nematode worm is not, but a mammal is).

Obviously, these issues are important to those concerned with animal
rights, or even with animal welfare. They should also be of concern to
feminists. I make that assertion partly because how science sees, or
constructs, images of nature has been intrinsic to feminist critiques of
science. Within that, human dominion over other creatures can be related
to the various other forms of domination that feminists criticize. But
there is also a more selfish reason for feminists to look carefully at how
science uses animals. It is, after all, scientific tales about non-human
animals that feed the fictions of biological determinism about women.

Notes

1 Extract quoted in Smith, Birke and Sadler (1994). In that research, extracts
  were coded rather than specifically identified, as we were concerned with
  general trends rather than wishing to single out individual papers or authors.
2 Donna Haraway, in *Primate Visions* (1989b) emphasizes the ways in which
  narratives of race, class and gender intersect in the construction of stories
  within twentieth-century primatology. In this sense, the construction of 'other'
  around the animals that are ostensibly the focus of study is also complexly
  related to, and interdependent with, the construction of 'other' human actors
  in the stories. Modern primatology is, she notes, partly a colonialist discourse,
  in which white scientists (often female) go to parts of Africa to study non-
  human primates. The animals in this are then situated not only with respect
  to the scientist, but also with respect to indigenous African human populations.

Part III

DEFINING ANIMALS

6

WINNERS IN
LIFE'S RACE?

In every group of animals there should be lowly forms which have
remained stationary while their fellows were becoming the winners
in life's race; just as we see the poor, the weak, the unenlightened,
and the unsuccessful living on beside the rich, the strong and the
highly cultivated and the successful. And these lowly forms should
retain features which are embryonic in the more highly developed
forms; just as the less favoured individuals among us are apt to be
childish and undeveloped.

(C. Lloyd Morgan 1885)

If our ideas of 'animals' in the wider culture are contingent and variable,
then should we perhaps turn to science for a clearer definition? It is
science, after all, that claims to pursue objective truth, and attempts to
define what species are; it is scientific theories of evolution which led to
Lloyd Morgan's belief that there are winners and losers in 'life's race'.
We can thus ask, what does science tell us about 'animals'? And what
light does this shed on the way that we think about the human/animal
dichotomy? Our hopes for clarification are likely, however, to be dis-
appointed. Conceptualizations of 'animals' or of specific kinds of animals
are as fuzzy and shifting – yet rooted in notions of eternal fixity – in
science as they are elsewhere.

Feminist biologists have criticized several areas of biological thinking,
and here I pick up on three of these that have particular relevance to
thinking about animals. In this chapter, I focus on ideas about evolution,
and about development; in the subsequent chapter, I turn to ethology,
the study of animal behaviour. Each raises concerns about how we
think both about gender and about humans/animals. In relation to
thinking about evolution, for example, feminist critics have explored

two important themes: one is the way in which static notions of species persist (despite the concepts of change implicit in ideas of evolution). The other is the way in which popular accounts of evolution have attended to the notion of 'survival of the fittest', or 'nature red in tooth and claw'. Apart from the gendered connotations explored in feminist critiques of science (Bleier 1984; Hubbard 1990), there are important implications for how we think about animals.

Feminists have not only been concerned with critiques, however; we have also looked at ways in which the thinking and practice of science might be improved if it took feminist claims more seriously. A secondary theme of these two chapters picks up on how these areas of study might change. Feminist criticism has largely been concerned with considering how science might change if it respected women's (or feminists') lives. But we can also ask what it might look like if it constructed different meanings about non-human lives.

In writing this chapter, I am aware that I may be setting up straw people to attack in making claims about what biologists say. To be sure, few biologists would be likely to make simplistic claims that, for example, genes cause everything. Yet they do sometimes use language that can lend itself to such interpretations. Scientists, as Evelyn Fox Keller reminds us, must inevitably be language users (Keller 1993). How that language is articulated, and used by others (the media, for instance), is part of the process of science. So, if I pick up on certain ideas about, say, how genes work or what 'animals' mean, then that is because these are ideas that have a popular currency. For that, it does not matter whether or not all (or even many) professional biologists employ them as directly or as simply as I describe them here. What matters is how those ideas gain a wider currency.

Evolving organisms

An important part of the history of biology has been the classification of organisms. Proper classification was necessary if order was to be perceived in the multitudes of different species (Ritvo 1991). Any system of classification seeks order through looking at similarities and differences; for centuries, that order was seen in terms of a linear and hierarchical chain which encapsulated both difference and similarity. It also encapsulated fixity; each species' place in the chain was held according to theological doctrine to be preordained by God.

Each organism, in this ranking, was similar to those immediately above and below it in the chain. Yet difference could be invoked by supposing a conceptual break; humans (or men) could thus be cast as different from the non-humans (or women) who came below. Feminists have, of course, been critical of this kind of ranking for its associations with gender, but difference of humans from non-humans can also be justified through supposing some kind of linear scale. Humans, according to this doctrine, were given dominion over other animals by God.

Despite Darwin's ideas about evolution, the supposition that organisms can be ranked in a scale of being persists; it goes hand in hand with the notion that evolution represents a linear progression towards perfection in the form of human beings. On this view, what separates us from other animals are particular traits (though we may argue over which ones) that can be found only in humans. This is not exactly what evolutionary theory says, but it is a common idea. Stephen Jay Gould notes the persistent iconography of the 'progress' idea in his book *Wonderful Life*; humans are depicted in popular accounts as somehow rising up from apes – the 'ascent of man'. 'The march of progress', comments Gould, 'is *the* canonical representation of evolution – the one picture immediately grasped and viscerally understood by all' (1991: 31).

There is a consistent failure in popular perceptions of evolution that ignores the multiple and interwoven branches of life's history. Somehow we always want to view the single twig (ours) as 'the acme of upward achievement, rather than the probably last gasp of a richer ancestry', notes Gould (35). This kind of thinking affects how we think about other species. Gould notes how, even when we depart from a linear progression and see evolution as represented by branches (a more accurate image), we still conflate placement in time (the bottom of the tree's trunk) with judgement of worth, so that 'our ordinary discourse about animals follows this iconography'. In other words, those animals we deem to be near the base of the trunk are lowly, primitive forms of life; the pinnacle is ourselves. Such conceits underlie how we think about humans, particularly in relation to non-humans. There may have been traits which species of hominids possessed uniquely, but as we are now the only hominid species left we cannot tell. The use of speech may have been one such trait. Equally, there may be traits which have arisen several times in evolution in different groups, just as the ability to fly has evolved in insects, birds and mammals. That is not to say, however, that these represent some kind of continuity or common ground between those species or groups. The belief in an evolutionary scale and linear progression towards the apex (mankind; *sic*) is precisely what underlies the human versus other animals thinking that I am criticizing.

The tension between similarity and difference is clearly central to my concern with the way we conceptualize animals; it is undoubtedly part of the complex ways in which animals metaphorically stand in relation to human society (Ritvo 1987; Tester 1991). In some ways, it is rooted in ideas of species that deny the sense of transformation implicit in

Darwin's thinking. The tension between similarity and difference rests on a peculiarly static concept of species.

The idea that species are fixed types, or essences, goes back to Plato, and was present in the development of systematic taxonomy of the eighteenth century. In principle, that idea was overturned by Darwin; biological species are communities of individuals that can interbreed, and whose characteristics are subject to continual modification by natural selection. There is, then, no such thing as an eternal form of a species. Yet static concepts of species persist, even among many biologists (Hull 1992).

From feminist perspectives, static ideas of other species inevitably feed into how we see nature and the human/animal relationship. Primarily, such ideas are heirs of the Platonic essence; the beast within is no particular animal, but the essence of animalness. Behaving 'like an animal' is to behave in presocial, uncivilized ways, or perhaps to allow our essential nature to surface.

Feminists have placed considerable emphasis on deconstructing the categories by which we have come to label the world; we can challenge the name 'woman', for instance, stressing instead the multiplicity of meanings involved (e.g. Riley 1988). Challenges to the unitary notion 'animal' (or nature) have been rather fewer, however; we continue implicitly to rely on concepts of nature as fixed, of non-humans as essentially *in* (asocial) nature. That, at the very least, ensures that we see non-human creatures as fundamentally (essentially?) different from us. Deconstruction relies on prioritizing human sociality, out of which we construct meanings; even the natural itself becomes a construction of the social (Fuss 1989). Possibilities of change are implied by social construction; constructions can always be remade. But despite the emphasis on deconstruction, non-human animals seem often to remain outside of feminist theorizing, mired in an unchanging, evolutionary backwater.

Yet evolutionary theory *is* primarily about change, about the possibility of *transformation* between species over time. It was this story of changes, combined with the way in which the idea of evolution emphasizes *continuity*, or similarity, between species that provoked controversy when Darwin's books first appeared (Rachels 1990). Somehow, though, the social sciences have tended to see fixity, rather than transformation, in non-human nature (Benton 1993), thus failing adequately to deconstruct the boundaries of 'nature'. It is no wonder that many feminists have sought to remove human behaviour from the realms of biology, if such a significant body of biological theory as evolution is believed to fix species forever.

Another major theme that feminists have often criticized in evolutionary theory is the interpretation of Darwin's idea of natural selection in terms of 'survival of the fittest', of overt and bloody competition. Because the phrase 'survival of the fittest' is well-known (and perpetuated through popular books and television), most people see evolution in those terms. The prevalence of the view that natural selection means competition among individual organisms may be a product – in part – of the language scientists have used to describe the processes of evolution. That language was inevitably shaped by the values of nineteenth-century Europe, but it persists today, and in turn helps to shape our beliefs about nature. One of the reasons why feminists tend to deplore any suggestion that we are like animals is because that carries an implication of savagery, of 'nature red in tooth and claw'. It is not caring, cooperative animals that come to mind when someone calls me a bitch.

Whatever the reasons for its centrality in nineteenth-century thinking, the idea of competition has remained all-embracing ever since. The use of the concept has, moreover, risen steeply in scientific writing in the last two decades (Keller 1993: 151–2). Competition is often interpreted as the struggle for existence, each individual pitted against all (though there are, in biological writing, many senses in which the word is used, as Keller points out).

Yet, several commentators have noted, there is plenty of evidence of mutuality in nature (Gross and Averill 1983). What is at issue here is not the extent to which we might (or might not) be able empirically to demonstrate such cooperation in nature. What is important is the overwhelming rhetorical importance of the idea of competition – so much so that the theory of natural selection is almost synonymous with a theory of competition to many biologists (Keller 1993). It is no wonder, then, that the two are confounded in popular writing.

Beliefs in 'nature red in tooth and claw' also colour how some people think about animal rights. Rosemary Rodd (1990) discusses, for example, the idea among some evolutionary biologists that morality might have a biological basis – at least in humans. She quotes, for example, one writer referring to the related notion that non-human animals inevitably 'act selfishly' because it is in their nature (or genes) to do so (1990: 212). Humans are somehow above such things by contrast.

The implicit appeal here is to a concept of evolution that, yet again, puts humans apart from other creatures, as the pinnacle of creation. Animals might be red in tooth and claw but rational humans (presumably male) are not, it would seem. More importantly, however, if only humans have morality, then we might concede duties towards animals, in accordance with our superior sense of morals, but we need not concede rights. Animals, to some, cannot have rights because they are too 'primitive' to understand anything but the amoral struggle for existence, the deadly competition of nature (e.g. Leahy 1991: 195).[1]

There is also controversy in evolutionary theory around the basic entities on which natural selection acts. As Keller has pointed out, much of modern evolutionary theory assumes that selection acts on individual entities (individual organisms or their genes; Keller 1993). What might be called the individualist programme in evolutionary theory illustrates, suggests Keller, how conventions of language can help to incorporate ideology; in this case, an ideology that sustains both competitiveness and individualism.

Keller (1993) notes that for any individual to reproduce (and therefore be reproductively 'fit') depends upon others, at least in species that reproduce sexually. No one can reproduce if there are not others who are also 'fit'. Reproductive fitness is not, then, solely a property of individuals, but is intrinsically part of the entire breeding population. Yet the language of evolutionary theory persists in giving priority to individuals, as though they can pass on their genes all by themselves. Competition creeps in here too, in the form of sexual selection; one sex (usually males) is said to compete for the limited reproductive resource (usually females). The language in which this is often couched remains as Victorian as when Darwin first wrote about it; promiscuous males and coy females still pepper the pages of biology textbooks (Hubbard 1990).

The language of evolutionary theory, then, has helped to construct images of gendered animals, the prototypes of gendered people. But what does that language – of competition, of natural selection, of individuals – mean for how we see animals? One obvious consequence of the language is that the concept of 'animal' is itself overlaid with the meanings ascribed to nature in Darwin's time. The animal is an individual, seemingly separated from others as it competes with them for resources. The problems with that are, first, that the language of individualism does not adequately describe all animals; not all animal kinds are as atomistic as that rhetoric implies. Some exist only as colonial forms (sponges, for example); the concept of the individual becomes quite meaningless for creatures such as these. Second, to see nature as composed of accumulations of individuals is to take as given the notion of individuals as the basic units of nature. Whatever else that is, it is an idea with a political context; it arose in a society espousing the tenets of liberal individualism. It also divorces organisms (individuals or populations) from their environment. At best, their environment is something that acts upon them from outside.

At another level, the individual and competitive emphases of modern evolutionary theory have reinforced a separation of self and other, which feeds back onto our ideas of ourselves as separate and different from other creatures. Nature, in this separation, becomes hostile, indifferent to human needs; a human is a solitary individual in an uncaring world (Keller 1993: 88). It is this linguistic construction that shapes our understanding of non-human animals. All too often, we simply fail to see interconnections between one organism and another (whatever their species), or between an organism and the ecosystem of which it is part.

Although evolutionary theory does emphasize transformation and change, and has tried to account for the existence of mutuality and cooperation (admittedly in terms of competition), its language continues to reflect beliefs in individuals and competition, and in species as fixed entities. I would not wish to infer from this that evolutionary biologists are unaware of these problems; indeed, much of the philosophical debate about species and evolution centres on the problems of language – what is meant by 'species', for example, or the problems of covert assumptions that what we mean is the evolution of vertebrates alone (Hull 1988). But the language used undoubtedly does have these biases, and that language becomes even less cautious when translated into popular accounts of natural history, which remains (for non-humans) bloody and instinctive. It is these more popular accounts – on television, radio, in magazines – that for most people, shape understandings of 'species', 'humans' or 'animals'.

Embryonic development

How embryos develop is a central question of biology. It is also one that feminists have explicitly addressed in our critiques of science, looking particularly at ideas about how embryos become sexed (e.g. Birke 1989; Fausto-Sterling 1992). Part of that critique has centred on the prevalent image of the developing organism unfolding from the information contained in its genes – an image that also has implications for the way that science helps to construct particular images of animals.

In the late twentieth century, following the discoveries of genes and DNA, we tend to think immediately of DNA as providing all the necessary information for development to proceed. Thus, the introduction of a book entitled *From Gene to Animal* begins: 'Living organisms are built up according to inherited instructions encoded in their DNA' (De Pomerai 1985). An individual animal thus becomes the product of the information revolution; if we learn to read the manual, we gain the secret of life.

Genetics claims to offer explanations of similarity and difference. All humans have genes in common, the argument goes, and so are similar. Similarity between mammals, too, depends upon genetic common ground. But differences also reflect the genes. The reason I am different from, yet similar to, both my parents has to do with the way I have inherited genes from both of them, as any elementary biology textbook will assert. What matters in this scenario is only what happens at the point of fertilization, when maternal and paternal genes mix. The rest of my development would seem to follow automatically from that.

Many feminist writers have noted one early theme in the history of studies of human development, namely the belief that only men's 'seed' (that is, sperm) gives form to the human fetus. The female uterus, in this view (which derived from Aristotle), supplies only nourishment. Aristotle agonized over the effect of women's relative 'coldness'; women, he reasoned, lacked the requisite amount of heat to transform menstrual blood into 'seed'. They cannot, therefore, form seed which would contribute to the form or shape of the developing fetus (Tuana 1989).

Whatever the speculation, quite how an egg became an embryo remained something of a mystery. One idea that began to take hold in the eighteenth century was the notion that embryos arose from an unfolding of material already present in the egg or sperm (a notion called preformationism). Thus each egg or sperm contained within it the material for all future generations, like a series of Russian dolls. Fertilization, according to this view, is merely a trigger, setting off the unfolding of preformed parts. The belief also led to flights of imagination; the microscopist Antony van Leeuwenhoek, for example, claimed in the seventeenth century to have seen a little man, the homunculus, inside the head of a sperm. There are apocryphal stories that other observers at the time made further claims, that they had even seen a homunculus taking off his coat!

By the end of the eighteenth century, however, an alternative theory of embryonic development was emerging. Rooted primarily in the experimentally-based laboratories of German universities, this focused on describing the ways in which embryos develop form from previously undifferentiated masses of tissue. In these traditions (called epigenesis), the emphasis was on the embryo gradually acquiring form – shaping itself, as it were. The different traditions became clear during the nineteenth century. The laboratory scientists focused largely on the mechanisms of fertilization and epigenesis through experimental means. Naturalists, by contrast, were more concerned with how characters are inherited (Farley 1982).

The determinism inherent in the preformationist notion of unfolding is clear; the environment of the embryo does little, and form emerges built into the sperm or egg from the start. Scientists may no longer believe in preformationism in just that way, but, as Susan Oyama has pointed out, the basic principles of preformationism persist in the way that we use ideas about genetics now (Oyama 1985). The notion of a 'genetic blueprint' for example is widespread, implying that the information contained in an animal's (or plant's) genes somehow unfolds into the developed animal.

Now what does all this have to do with how we see animals? Feminists, among others, have certainly been critical of the way in which such ideas lend themselves to fixed visions of human potential. But they do so about animals, too. The sense that we get, of an organism simply emerging from the drawing board, denies the organism agency, reducing it to a set of blind instincts that are wired into the animal at conception. There is, I think, a very strong tendency for people in western culture to believe precisely that about other species; 'animals' are little more than bundles of instincts.

Indeed, so dominant is the view that all that it takes to make an animal is encoded in the genes that, by 1986, geneticists were claiming that the study of embryology was subsumed into the wider field of genetics. Making this point, Fausto-Sterling (1989) goes on to note that embryology – which has emphasized epigenetic development – has been locked in a power struggle with the emerging discipline of genetics for most of this century. The persistence of preformationist ideas in the 'blueprint of the genes' is, she argues, significant because it implies that the context in which an organism develops is largely irrelevant. 'If the code *is* the organism', she notes, 'then it should be possible to gain power over the developmental process by removing it from its original context and placing it in the laboratory' (1989: 321). That, indeed, is a major theme of much research in genetics; the genes' context (the fertilized egg, say), the embryo's context (in a mammal, travelling along the Fallopian tube to become embedded in the uterine wall), the fetus's context (attached to its mother's uterus via the placenta, hearing her heartbeat, and so on) – all these become irrelevant compared to the all-powerful genes.

The alternative tradition is one that feminists have emphasized, and which sees organisms in a more epigenetic light – that is, as gradually developing in the context of (and in interaction with) the immediate environment. In this picture, the organism is actively engaged in a process, rather than blindly driven. Here, the causes of developmental processes are multiple. Significantly, they include interactions with the environment – be that the ocean or a female mammal's womb.

An example of how these developmental processes are ignored is provided by some accounts of sex determination (by that I mean the process by which an individual organism becomes anatomically female or male). Popularized accounts of sex determination (in biology textbooks, for example) refer to the sex chromosomes (X and Y) as *the* primary locus of sex determination. More recent research has suggested that a 'master-switch' on the Y chromosome is basic to switching on male development (in its absence, on this theory, femaleness develops).

The focus on chromosomes and genes overemphasizes the dichotomy, male versus female. Just as human gender, as socially constructed, is far more complex than a simple dualism, so too is sex as the construction of biological bodies. Nature does produce many examples of intersex. There are also many non-human species in which the sex of offspring is *not* determined by the possession of particular chromosomes or genes; rather it is the product of changes in the environment. Temperature, for example, may alter development and determine the sex of offspring in many species (Farley 1982); some species of fish seem to change sex with changing social conditions. The effects of hormones during development are also an important, and complex, factor (a point to which I return in a later chapter).

Another assumption informing genetics is that relative uniformity or similarity must be the direct result of gene action. If different creatures have some features of their structure in common, the reasoning goes, then this must be because they share genes. How structural differences are explained depends upon how they are categorized; some are attributed to environment, some to genes. If they are differences between species, then the assumption will usually be that they reflect genetic, interspecies, differences; if they are variations within a species, then they could be due to genes (a genetic mutation, for example) and/or environment. But similarity tends to be explained primarily by recourse to genes. I am simplifying the argument here, and few biologists would make that claim quite so simply. Yet the general drift of such reasoning is sufficiently strong for these assumptions to find their way into popular accounts and elementary textbooks.

Similarity of form, however, need not be the result of possessing certain genes. An alternative, epigenetic, view is that constraints could arise in the development of structural form – mechanical constraints, for instance – that themselves constrain how cells behave around them. In the early development of embryos, cells migrate often long distances along particular paths, but what determines those paths may be mechanical structures or electrical fields (Goodwin 1984). Of course genes are involved in all this; the initial development of those mechanical structures is likely to involve particular genes. But the widespread assumption that genes *directly* cause the emergent patterns can be challenged.

The prevailing emphasis on unfolding from the genes, combined with reductionist accounts of sex determination, have implications for how we see non-human animals. Western culture tends to believe that non-human animals are the products of a specifically determined, or genetic, heritage. Ideas of other animals are thus set up as fixed, locked into our (equally constructed) ideas of what genes are. Obviously, there are wide differences between species; the behaviour of many invertebrate animals does indeed seem to be relatively fixed and unvarying, while that of many vertebrate animals – especially primates – is not. But the predominant notion is indeed one of the 'hard-wired' animal, the beast-machine.

Moreover, if we see the development of 'sex' differences in animal behaviour as fixed by genes or hormones, then we remain trapped into dualistic thinking. To infer that (say) the development of rat behaviour is fixed by hormones but that of humans is flexible is to impose a chasm between humans and other animals. This can backfire, as I shall argue in later chapters, simply because there will always be people who will claim some similarity. It is but a small step, then, to talk about the fixity and biological determination of human behaviour by analogy.

Finally, emphasizing the role of genes in development helps to reify them, one result of which has been the massive injection of funds into the Human Genome Project, the attempt to map every human gene. The eugenic implications of this research, and of its underlying assumptions, have been dealt with elsewhere (e.g. Hubbard 1990; Hubbard and Wald 1993). Here, I want to emphasize two other points. The first is that reifying genes feeds back into the process of decontextualizing them. Partly this is because it ignores the complex interactions between genes themselves. As Hubbard and Wald have noted, despite the extravagant promises accompanying advertising for the Human Genome Project, mapping the genes will not tell us where exactly to find the 'genetic inheritance' of humankind (1993: 158–9). Partly, too, reifying them allows us to see *only* the genes, devoid of their physiological context, the organism itself and *its* environment. In her analyses of language and science, Keller notes how, increasingly, genes are

> presented not only as having no need of the bodies in which they are housed for the processes of 'reading' and 'interpretation'; they no longer need even their own chemical structures for their existence. The substantive component of the gene is said to lie in its nucleotide sequence, and that can be stored in data banks and transmitted by electronic mail.
>
> (1993: 179–80)

Reification, indeed.

Transgenic organisms are a good example of how the effects of genes can be taken apart from their context. Scientists can produce sheep, for example, with human genes to turn them into animal factories to produce molecules useful to human medicine. But what worries many critics is that this kind of use (or abuse) of non-human animals ignores what happens to the inserted gene *apart from* its attributed role of molecule-factory. How does it interact with the sheep's genes? Or with its cells? Or with its wider environment? And what are the consequences of those interactions?

While species differences are seen as encoded in the genes, DNA is just DNA. We can contemplate putting a 'human gene' into a sheep just because we tend to see quantities of genes as embodying the essence of the organism involved. Given enough 'sheep genes', the sheepness of its nature can be preserved and there is no threat to species boundaries. Yet it is precisely because DNA is DNA (that is, there are small differences in its structure between related species such as two different mammals) that allows us to put a human gene into the sheep genome and expect the sheep's cells to understand the instructions. But if you start to think about the similarity of DNA, or the similarities of its function, then there is little difference between sheep and human; the boundaries start to dissolve. To avoid that worrying prospect, we can label the genes as embodying essence. *Human* genes must have something special about them, an essence of humanness, that gives them a territorial boundary to preserve them against the essence of sheep.

It is not just that, in current theory, development has become encoded in the genes, thus ignoring the complex processes of development. What has also happened is the creation of 'the gene' as a representation. What it embodies is a particular body form or type (the alleged 'homosexual gene' springs to mind). Genetics as a system of representation both challenges concepts of species as fixed (in the practice of, for example, transgenics) and reinforces them (by incorporating notions of essence). Genetics thus presents to us categories that are fracturing and shifting across boundaries that we are accustomed to call species. DNA may be 'just' DNA; but when what that means is that the boundary between humans and beasts comes under threat then we get very worried indeed.

The predominant framework of genetic explanations encourages belief in organisms as *just* their genes, however. We might anxiously add human culture and society to how we see ourselves; but non-humans remain determined by their DNA. The rapid growth of molecular biology in the twentieth century – and particularly the expansion of genetics and DNA technology in recent decades – fuel that image of animals as puppets of their genes.

Yet, just as feminist critics of science have placed more emphasis on the context of development, so too can we stress context for other creatures. Embryos and young infants, of whatever species, are better thought of as active agents; they not only become, but they also make themselves. I am reminded here of Simone de Beauvoir's well-known phrase, that woman is made and not born. The young animal is similarly not 'born' as the product of nothing but its DNA, but made. To be sure, its genes and physiology may be part of that making; but so too is its physical and social environment – and its own 'making over' of that environment.

As I thought about this, I wondered again about learning and training in the development of young companion animals – dogs, cats or horses, for instance. Clearly, there are species and breed characteristics that, presumably, are influenced by the genes: my lurchers, true to their greyhound parentage, love to run. My cairn terrier is happiest pretending she can catch the rats she stalks (even if the rats are only too well aware of her presence and have gone elsewhere). When a puppy is born, however, it is likely to be socialized both into canine sociality *and* into human sociality. It becomes not 'essence of dog' but a creature fulfilling a role we ascribe to it.

Canine and human expectations inevitably help to shape those processes of socialization. Even the 'breed characteristics' may not simply be due to genes, for how much is the cairn terrier responding to my beliefs about 'what terriers do'? How much is the lurchers' obvious enjoyment of running influenced by my expectations of such athletic-looking creatures? We understand too little of how the human/animal social environment interacts with whatever inheritable characteristics we have selected for in generations of breeding domestic animals. How animals develop must depend on much more than just the genes.

Note

1 The concept of rights is problematic; it does not, however, necessarily depend on moral understanding. Notions of human rights are, for instance, applied to small children or even fetuses, which are clearly not (yet) capable of such understanding. In the case of severely brain-damaged children, moral understanding may never be achieved.

7

BEHAVING DIFFERENTLY: ANIMAL ABILITIES

A couple of days later it was my turn to watch the cows to see who needed to be inseminated . . . I studied her closely for any signs that might indicate why she was losing her milk . . . [then] I saw the real Henderson and I understood . . . She wasn't a dairy cow – she was a physicist trapped in the body of a cow. She was a scientist and she was supposed to be wearing black horn rimmed glasses and a pocket calculator . . . Renee Richards never had it as bad as Henderson did, standing there, ankle deep in cow shit, wanting only to study a table of logarithms and instead having to endure three milkings a day for the rest of her life.

(Rose 1988)

How animals behave is one important source of information for us in understanding them; Henderson kicked and refused to be milked, and the narrator had to find an explanation. But more important are our *beliefs* about how animals behave, what they are capable of doing, and the meanings we attribute to that. We smile at Henderson's story because it seems ludicrous to us to suppose that a dairy cow could aspire to being a physicist.

It is those beliefs about what animals can or cannot do that are marshalled in any argument about whether or not other species possess abilities akin to ours. What we know (or think we know) about other animals may come from many sources: personal experience, anecdotal stories, training animals, fiction, or scientific accounts, for example. But in any debate about whether animals have intelligence, say, what really counts is data that have been acquired through a scientific study, an experiment.

Yet anecdotes abound that suggest that some animals might have quite special abilities that would often be missed by standard 'scientific' approaches. Speak to anyone who has worked with assistance animals, for instance, and they will tell you countless tales of the dog that leapt behind the wheels of its owners' wheelchair to stop the chair from running backwards; or the dog that succeeded where humans had failed in awakening a child in a deep coma by licking her face; or the dogs trained to work with people with paraplegia, picking things up for them and opening doors.

In this chapter, I want to look critically at what kinds of things about animals the scientific studies can, and cannot, tell us. I begin this exploration with a look at some of the limitations to understanding that result from the reductionism of scientific methods. To be sure, scientific approaches have yielded a wealth of information about different species, perhaps including our own. But there are limits to what can be achieved; focusing on groups or species, for instance, may blind us to the richness of individual experiences.

The rest of the chapter takes up two behavioural questions which are relevant to thinking about humans versus other animals. These are: the question of the extent to which non-humans feel pain or may suffer as we do; and the question of animal intelligence and consciousness. These topics are ones that raise questions for me about how we think about animals versus humans, and also about how we think about gender – themes that I take up in the last section of the book.

Studying the behaviour of animals

Ethology (the study of animal behaviour) has been much criticized in recent years for its apparent support of crass biological determinism; certainly, there have been many writers who have appealed to ethology to provide support for their views on, for example, human aggression or male dominance. These traits are said to be due to instinct rather than learning; the assumption is that if this is so then they are fundamentally ineradicable. Not surprisingly, this has met with criticism (e.g. Bleier 1984; BSSRS Sociobiology Group 1984).

Ironically, developments in ethology have also helped to provide justification for those who wish to challenge the human/animal boundary. Contemporary ethology has two main roots: one strand, particularly in Europe, was part of a natural history tradition, that focused mainly on the lives of wild animals. The other root (more evident, perhaps, in North America) derived from behaviourism, with its emphasis on experimental studies of how animals learn. Behaviourism, in its turn, was a reaction at the end of the nineteenth century against the use of mentalist constructs, like 'mind' (which could not, by definition, be studied scientifically). So, for much of this century, experimental studies of animal behaviour explicitly denied any kind of 'mind' to animals (and only begrudgingly, if at all, allowed it to humans; see Mackenzie 1977, for an account of behaviourism and its background).

European ethology was also not concerned with 'mind', but focused instead on what was at the time termed instinctive behaviour. Writing in the 1950s, Niko Tinbergen emphasized that to approach the study of animal behaviour we need to ask four kinds of questions. These are:

1 What are the immediate causes of the behaviour? This could be an external stimulus (such as a robin attacking a bunch of red feathers), or an internal one, such as rising hormone levels.
2 How does the behaviour develop within individuals? We might ask, thus, how play behaviour develops in infant monkeys.
3 How did the behaviour evolve?
4 What function does the behaviour serve? Laying down food caches, for instance, serves the function of providing a squirrel with food in the winter.

It is the last of these that has developed strongly in recent years in the form of sociobiology – indeed, so much so that Barlow (1989: 2) began an article with the provocative claim that 'Ethology is dead, or at least senescent' in its classical form. The fashion for sociobiology (or behavioural ecology as it is also called) has meant that the other kinds of questions have perhaps been neglected, as Barlow points out.

Current understanding of why animals behave as they do is, as a result of that fashion, filtered through assumptions that animals are efficient, functional entities, or perhaps even automata. It is, moreover, an understanding largely constructed within specific cultural contexts, as Donna Haraway reminds us. Could such optimally efficient machine/creatures have been discovered by any other culture but that of the West? ·

Barlow is quite right that, within behavioural ecology, there is an emphasis on the flexibility of animal behaviour; he cites as an example the role of learning in foraging behaviour. But seeing animals as optimally efficient is to describe them in a language which itself structures our perceptions; the language of sociobiology is, as many critics have pointed out, one that represents animals as driven by genetic or evolutionary imperatives (albeit ones that allow for some learning). Ironically, it is also one that is anthropomorphic, encouraging us to think that animals are strategists, in the same kind of way that humans can be (Kennedy 1992). The language of science, then, is the first way in which our understanding of animals is structured. Increasingly, this kind of language is one that informs popular understandings through television and magazine articles.

The methods of science

The methods used in science help further to structure the understanding of animals that emerges. Despite its roots in observations of natural history, the formal study of animal behaviour has largely operated within strictly defined methodological limits – that is, reductionism. It is much more straightforward to design an experiment looking at simple variables (altering hormone levels in an infant animal, for example, then seeing what happens to its adult behaviour, or training it to press levers for rewards), than to design one allowing for the interplay of many complex variables. As a result, what we can learn about animal behaviour is limited by the constraints of the methodology.

One consequence of emphasizing the experimental, scientific aspects of behavioural study has been an increasing emphasis on *statistical* norms. Where once it was acceptable to write about one's own observations of specific animals or phenomena (as indeed Darwin did), by the middle of this century articles in animal psychology emphasized hypothesis testing and quantitative analyses. Both the individual animal, and the individual, observing scientist, have become subordinated to the abstractions and numbers of the scientific text (Bazerman 1988).

In turn, an effect of this emphasis has been that individuals have largely been ignored (with the notable exception of specific studies of individual animals' learning capabilities, such as the well-known studies of sign language acquisition by chimpanzees; Premack and Premack 1983). One reason for this is the assumption that the appropriate unit of study in biology is the species.

The assumption that underwrites studies of animal behaviour, by contrast, is that the groups being studied represent the species – and hence the norm. 'Deviant' behaviour of individuals has not, on the whole, been seen as a relevant problem. In that sense at least, *what we know* about animals from science implies that they have less individuality than we do, but what we know from scientific studies has been shaped by a methodology that makes many assumptions about normality and the 'typicality' of certain patterns of behaviour.

From natural history, we learn much about the ways in which particular species are adapted to particular habitats; from the experimental traditions of ethology, we learn for example about why and how animals choose to perform a particular behaviour in particular contexts. Yet individual histories are downplayed in both; the animals studied stand to represent a category we call the species. It is that representation that matters in the scientific narratives; the individual merely stands for something else.

If scientific methods reduce our awareness of individuality in animals, then how we perceive other animals will be shaped by their role as signifiers of 'species'. Yet that is not the only way in which we 'know' animals. We might also be able to generalize about specific species from having understanding of individuals, as animal trainers do. Trainers, however, might be willing to grant to animals qualities such as subjectivity, that scientists would tend to abhor.

Many people in western culture have had some experience with companion animals that would make them inclined to believe in animal subjectivity and capabilities. In some ways, that is a problem for the scientific accounts, which often seem to be at odds with what we might call common-sense notions of what animals can or cannot do. Part of the problem, though, is that common-sense notions vary. In teaching courses about animal behaviour that focus on wild animals, I typically encounter the belief that animal behaviour is instinctive – the 'hard-wired' animal of genetic drives, somewhat analogous to a living computer. Yet the same people may also concede that the companion animals who share their lives are not so hard-wired; we believe that these animals have choices or feelings and enter into our lives.

The distinction, I suspect, owes much to how we learn about animals. Increasingly, learning about 'wild' animals in nature is something we do from the couch in front of the TV. The creatures populating these images seem innately to follow preset patterns of behaviour; they are always representing the concept 'penguin', or 'walrus' or whatever. Living with companion animals, however, gives us a different set of experiences; here, we bring them into our social world, and project onto them our own emotional expectations. To these animals, then, we might grant agency and intelligence. By contrast, they do not represent concepts of species. My companions at home are Ginny, Lisa, Penny and Tess; they do not typify 'the dog' for me (though they may do that for others – perhaps especially if the other is threatening to attack me or burgle the house!).

Pain and suffering

That animals can suffer and feel pain is obvious to many people, and is one important reason why the anti-vivisectionist cause has appeal. Causing suffering, some would say, is always morally wrong; it can never be justified, not even by recourse to claims about the usefulness of the experiments. Others, however, claim that some human uses of animals, including science, can sometimes be justified, provided that the gains in knowledge are demonstrable, and that the suffering is minimal. But that is the crucial question: how can we determine whether pain and suffering are 'minimal'? What does it mean to say that an animal that is quite unlike us (a snake, say) 'feels pain'?

Whatever its limitations to understanding individual experience, ethology can be useful in debates about whether, or to what extent, animals feel pain or suffer. Some writers are so keen to avoid attributing capabilities to animals that they deny that non-humans can truly feel pain; to experience pain as humans do, the argument goes, requires a conscious awareness of the consequences (Harrison 1989; Leahy 1991). Only humans can thus fully experience pain, because only humans have such awareness. On this line of thought, animals may be able to react to noxious stimuli, but they cannot understand the meaning of pain. Thus they cannot experience it, and their pain matters less than would ours.

What kind of evidence would be acceptable to show that non-human animals can feel pain or suffer? Most people take the line that animals can feel pain; that, after all, is why anaesthetics are used during animal surgery. Few dog owners are likely to agree to having Fido fully conscious throughout his operation. This common-sense view is based on two sources of evidence; first is the assumption of similarity between us and other species of vertebrates. So, if anaesthesia works for us, then it is likely to work for species having basically similar nervous systems.[1]

The second is that at least some species of injured mammals may whimper or squeal, or otherwise behave in ways that suggest discomfort at least. This criterion can be tricky to use, however, as some species, particularly ungulates (hoofed mammals) do not usually whimper and squeal; how would we know when Henderson is in pain, for instance (see Bateson *et al.* 1992)? Nonetheless, squealing is a response that most of us would judge to indicate pain in at least some kinds of animal. Yet should we judge the ability of other species to feel pain by such an anthropocentric criterion as the similarity of their nervous system to ours? Another way of looking at distinctions between species might be to focus instead on nervous system complexity. On this basis, we might feel that at least some invertebrates – the cephalopod molluscs (octopus and squid) for example – should be given the benefit of the doubt (as indeed some people have argued).

'Suffering' is even harder to infer. Debates about suffering often tend to rely on the assumptions either that we would indubitably suffer in similar circumstances, or the animal must be suffering because (say) it is deprived of the opportunity to do something that 'it would normally do in the wild'. Yet neither is easy to adduce for another person, let alone for an individual of another species. Bateson *et al.* (1992) point out, however, that we do make assumptions about the pain or suffering of other humans; we infer it from their appearance, or the cries they utter, or simply because they say so. Surely, then, we should do this for at least some other animals? We can do this, as we do for fellow humans, on the basis of seeking comparable mechanisms or behaviour. One practical example of such an approach comes from the assessment of behavioural disturbance in laboratory animals in response to routine procedures; for example, giving rats a saline injection makes them more active in proportion to the dose of salt solution they received, while their activity in response to being picked up by an inexperienced person can be lessened if that person wears gloves (Barclay *et al.* 1988). So, laboratory workers could use measures such as these to assess possible suffering. This does not, of course, tell us whether the animals perceived pain as we do (so it would not convince the sceptics); we may never know that.

But surely we should give them the benefit of the doubt? After all, the rats' reactions are similar to what I would expect of myself; if I received an injection into the sole of my foot, I have no doubt I would be 'disturbed' for days afterwards, just as the rats were. The only difference is that the rats cannot speak.

Animal capabilities?

Speaking is one thing that humans alone can do (as far as we know) – and to which some commentators attach great significance. The argument here is that using language has given humans an evolutionary edge, allowing us to develop an intelligence that is not only superior but different in kind. Much of the debate about the possibility or otherwise of 'animal minds' has, in recent years, focused on such things as tool use, or ability to understand language. Brain size or complexity, too, is often invoked; humans, inevitably, are deemed to have the most complex brain structure. The debate is often polarized; on one side, there are those (like me) who believe that we in science have underestimated animal abilities. On the other, there are those who believe that humans alone possess rationality and language, and that there is an enormous chasm, therefore, between us and any other species.

Cutting across these debates are the concerns of many scientists to avoid anthropomorphism (the projection onto animals of human traits; see Kennedy 1992) or teleology (the assumption of purposiveness in nature; see McFarland 1989). The claim here is that we do not need such obfuscating assumptions as animal minds; simpler, more mechanistic, descriptions of how animals work will do. Others are more willing to see intentions in animals, or to anthropomorphize about an animal grieving over the death of a companion.

A primary problem in much of this debate is how to *interpret* the findings. Consider, for example, the story of Clever Hans, a horse who was believed to be able to count. This story is cited in many accounts of animal capabilities, as a warning to those who would infer such abilities as counting. If asked how many objects there were before him,

Hans would tap his foot until the right number was reached. Then someone discovered that the trainer was giving inadvertent cues, such as slight body movements or intake of breath. If the horse was accompanied by a person who could not count, then he did not know when to stop (see Carruthers 1986: 235).

Now this tale certainly argues against any ability to count in the sense we might use it; we are supposed to infer that the horse did not possess any such cleverness. That, indeed, is the context in which the tale is typically told in textbooks. But is this not rather arrogant? In the first place, there is now evidence that some species may be able to 'count' (Davis and Memmott 1982). More importantly, to tell the tale as an example of animal stupidity is to ignore the rather remarkable abilities of the horse to learn to respond selectively to very slight cues from the trainer. I doubt if I could do that, even though I have acquired the human skill of counting.

In recent years, a renewed interest in 'animal minds' has developed (e.g. Walker 1983; Griffin 1992). Cognitive ethology, to give it a name, has revealed a wealth of ways in which non-human animals use complex mental concepts, or have consciousness, ways in which they might have a moral sense, or ways in which they can 'lie' and so manipulate the behaviour of others. There is plenty of evidence, for instance, of abilities to solve problems among animals. In one study of captive ravens, the birds quickly learned to use beak and feet to get at a piece of meat hanging suspended from a long string (which was otherwise inaccessible); in another study, ravens learned to locate food according to the number of spots (from one to seven) on the lid (see Griffin 1992: 104, 136).

Deception, too, seems to be widespread in nature. At its simplest, this involves mimicry, perhaps unconsciously; we cannot know, for example, how aware the plover is of what she is doing when she feigns a broken wing to lead predators away from the nest. Vervet monkeys giving 'fake' alarm calls when another monkey troupe comes near (and from which they flee) seems a clearer example, particularly since the call-giver often follows this up by wandering, apparently unconcerned, into the open space vacated by the other troupe (Cheyney and Seyfarth 1990). The heritage of behaviourism would have us deny consciousness on the part of these animals; on the other hand, as Griffin notes (1992 57), 'we cannot tell, but perhaps it is best to keep an open mind and not dismiss such possibilities out of hand'.

Indeed, those who train animals might wonder why it has taken science so long to catch up with what they have long known about animal thinking. They might sometimes adopt the languages of science – talking behaviouristically of conditioning, for example – while simultaneously believing in the animal's abilities to form concepts (Hearne 1987). Admittedly, the kinds of animals that we train in depth are nearly always mammals or birds; we know relatively little about the concept formation of other kinds of animals.

A central problem with the human versus other animals polarity is that it ignores the obvious point that all species are different. It is not just that humans are different from 'other animals'; 'every [kind of] animal is the smartest' if you know how to ask questions of its intelligence that are appropriate to its way of life rather than questions dictated by beliefs in an underlying animal stupidity (Bailey 1986). Animal trainers often know this; we ask dogs (a predator species) to chase balls, but we would think it foolish to ask a horse to do so.

Yet interpreting 'stupidity' is not easy, even among ourselves. In their analysis of the meanings of 'animal consciousness', Radner and Radner (1989) note the interpretations given to examples of alleged animal automatons. They cite, for example, the case of a species of bee that was fooled by experimenters into repeating a particular behaviour pattern over and over again (the bees respond to the odour of oleic acid as an indicator that there is a dead bee in the hive to be removed. The experimenters daubed a live bee with oleic acid, and found that the bees repeatedly tried to remove it). Now the behaviour can be thought of as illustrative of bees failing to recognize a problem, a logical mismatch. But, as the Radners point out, why are we so sure that they *are* simply being stupid?

The Radners note, for example, our own uncertainties about how to recognize death, even with the aid of high-tech medical apparatus. More importantly, we make allowances for humans to be credulous or gullible even when they persist in irrational beliefs, while

> animals . . . are expected to be perfect little scientists. In order to earn the epithet 'conscious' they must be proficient in logic, ever ready to change their beliefs in the face of available evidence, careful to take all considerations into account. When people fail to live up to this idea, we say they are all too human. When animals fail, they are said to be machine-like.
>
> (Radner and Radner 1989: 180–1. Also see Griffin 1992)

Similar problems arise when we begin to think about language. This, according to many writers, is not only the basis of human culture, but is unique to our species; it is language that sets us apart. I would not wish to deny that humans *seem* to be the only species using spoken language, with all its richness and syntactical structure. But there are problems nonetheless. In the first place, it is difficult to know just what is happening in the communication of non-human animals. Much of it is not auditory, thus I doubt very much that I can have any inkling of the largely olfactory world of a dog, let alone of the mental states that correspond to it. What would it mean, for instance, to be one of the bloodhounds used to search for missing people, on the basis of brief exposure to a faint scent? And even among those species that use complex auditory signals, such as whales and dolphins, it is difficult to interpret those signals.

Griffin (1992) notes another problem inherent in the emphasis placed on syntax as critical for spoken language; this emphasis, he suggests, has been 'enunciated primarily by scientists and scholars whose native language is English, one of the few human languages where word order provides almost all of the syntax' (217). Other languages convey meaning through inflection. Perhaps the use of inflection would be a more productive way to study animals' ability to understand language, Griffin argues, but no one has yet tried.

Difficulty aside, an underlying assumption of much research on animal communication, and on whether apes can learn sign language, is that we can legitimately ask the animal whether it understands *our* communication. That is the acid test of animal abilities, it would seem. Yet this is not comparing like with like; in teaching sign language to animals, we forget that this will be a 'second language' for them. As Lesley Rogers has pointed out, the studies should make comparisons with humans who learned signing as a second language *and* should begin by 'assuming that their own species-specific patterns of communication are their first language' (1995). How often do studies of animals and language start from the existing abilities and communication of that species?

We might ask, too, why people do feel worried by the notion that some animals might cross the language barrier. Ursula LeGuin explores this theme in her fictional account, 'The Author of the Acacia Seeds'. She notes in a preface that

> Some linguists deny the capacity of apes to talk in quite the same spirit in which their intellectual forebears denied the capacity of women to think. If these great men are threatened by Koko the gorilla speaking a little [sign language], how would they feel reading a lab report written by the rat?
>
> (1987c: 157)

The story is written from the perspective of 'therolinguists' of a future time, who have learned to decipher animal languages, such as Penguin, or Dolphin. One speculates whether it will one day be possible to read the languages of plants, reminding readers that 'so late as the mid-twentieth century, most scientists, and many artists, did not believe that even Dolphin would ever be comprehensible to the human brain – or worth comprehending!' (LeGuin 1987c: 175).

Scientific demonstrations of animal problem-solving abilities pose a problem for those determined to defend human superiority. Leahy (1991), for example, concedes that other animals have some ability to solve simple problems, but retreats into language as the critical divide. He cites the example of work with chimpanzees learning to use symbols through computer keyboards, and then learning to use this to communicate with each other. This, he admits, is a 'methodological breakthrough'. But, lest we start to think how clever the animals are, he asserts that

> it is a far cry from the use of language, observable in very young children, in which the exchange of information rapidly becomes an end in itself, rather than a means of problem solving . . . which we know to exist in animals anyway.
>
> (Leahy 1991: 163)

That kind of claim assumes that thinking is synonymous with language, that other ways of thinking are impossible to envisage. Walker (1983: 377), reviewing literature on animal thinking, notes how often critics assume that being non-verbal means that animals cannot think, pointing out that 'adults, and especially academic adults, are often reluctant to accept that their own conscious thought is possible without the internal use of language'.

There does seem to be a tendency in the literature on human versus animal abilities to see language and rationality as the archetypal bases of 'thought'. Insight and intuition do not, it would seem, feature here. Rationality is, of course, not only constructed as uniquely human; it is also masculine in the traditions of the West – the 'man of reason' (Lloyd 1984). Human, in these traditions, becomes not only cast as different from animal, but synonymous with man (and thus apart from woman).

The use of tools is another criterion sometimes applied to separate humanity, on the basis of the claim that all humans use tools, whereas not all other kinds of primates do so (see Candland 1987). Yet a hidden assumption here is that human cultures all have the same kind of tool use; this is the standard against which allegedly the sporadic use of tools by other species is compared. Perhaps they do, now that we live in an age of jet travel and fast communication, but did they in early hominid history? McGrew (1992) notes that anthropologists happily study variation in human cultures, while simultaneously defining it as uniquely human. On those grounds, variation within other, non-human, species would not count in these definitions as 'cultural'.

Yet surely differences between groups are just what counts as 'culture' – even for animals. McGrew cites the example of group (cultural) differences between different populations of chimps living in Africa; one population may preferentially eat foodstuffs that another population will not, for example. McGrew points out that if such differences were noted between human populations, we would assume them to be cultural:

No one's first explanation of (for example) a person's avoidance of touching food with the left hand would be that it is natural. Yet what would we make of exactly the same behavioural pattern if shown by apes? . . . suppose we found a population of previously unknown hominids who avoided contact with surface water. Not only did they not swim in lakes, paddle in ponds, or ford rivers, but they never waded across the shallowest and narrowest streams and even detoured around puddles on paths. If we perceived of these creatures as human-like, we might try to explain the act in terms of custom, tradition, ritual, or even symbolic taboo. If we perceived of them as ape-like, we might think of the behaviour as instinctive, hard-wired, species-typical, adaptive, etc. In fact, the chimpanzees at Gombe show just this reticence . . . while chimpanzees at Kasoje do not.

(1992: 217)

Animals' different experiences are typically relegated to some feature of their biology, their intrinsic nature; for humans, by contrast, we see explanation in terms of cultural difference.

A similar tendency to apply different criteria to animals and humans (and then – surprise! – to conclude that animals cannot do as well as us) can be found in claims that rely on brain size/shape. Feminists, among others, have been critical of claims about the alleged inferiority of women's brains (or those of non-white races) on the basis of brain size, not least because such differences disappear when body size is taken into account (Genova 1989). Yet crude measures of brain size are seen as a fundamental criterion of intellectual capabilities when humans are compared to non-humans.

Like comparisons between humans, these should (but do not always) take body size/shape into account. They should, too, compare like with like. The brain structure of birds is quite different, for example, from that of mammals in which the particular structures of the neocortex have developed. But more importantly, any comparisons should allow for the circumstances in which the non-humans have lived. Brain structure and cognitive abilities can vary enormously *within* a species in relation to experience; the thickness of the brain cortex, for instance, in rats that have been exposed to enriched, complex environments is much greater than that of rats living in impoverished environments. Yet what we know about the brains of non-human species typically comes from animals living in restricted environments – quite unlike the humans to whom they are routinely compared (Rogers 1995). It is no wonder that we are willing to grant some intelligence to our companion animals, who share the complexity of our daily lives, while we might deny it to the impoverished rat living in a laboratory cage. And her deprived life must inevitably produce a brain that will seem defective compared to our own.

Bringing animals into (scientific) culture

Perhaps the most important theme to draw out of this and the previous chapter is that, even within the development of scientific biology, there are different, and often countervailing, tendencies in the way that 'animals' are portrayed. We can see non-humans as being blindly driven by instincts, genes, or whatever; or we can see them more in terms of being self-directed and having 'minds of their own'. These alternative views have different implications for how we think about non-human species, and thus about how we think about ourselves. To see, for example, other animals as hard-wired and determined by their biology leaves us with only two options for ourselves. We can either claim to be equally hard-wired, or we can espouse a profound discontinuity and assume that we are different from other species. As later chapters will explore, this either/or position is simply untenable, and creates problems for feminist theorizing.

Alternatively, we can place more weight on animals as self-actualizing and developing. Again, this is not to deny that some species of animals may still appear to be quite fixed in their behaviour; in some cases, a more 'hard-wired' account may indeed be appropriate for the behaviour of, say, a slug. We might instead *begin* from the assumption that an animal is less determined. This stance poses quite different questions for how we see ourselves; assuming evolutionary continuity, for example, immediately becomes less threatening to our sense of our own achievements. Rather than 'reducing us to the bestial', it begins to elevate other species above that imaginary level.

An underlying thread to all this is the close linkages between how ideas about animals are constructed and used in science, and how they are constructed and used in the wider society. There is a reciprocal relationship; how animal societies are seen within scientific accounts both reflects and reproduces assumptions we make about our own society. How science has looked at animals thus becomes crucial. Not only do they potentially mirror us, but if science had traditionally downplayed their individuality, then that mirroring can begin to threaten ours. And if science has traditionally denied their self-consciousness, so too does it threaten ours. It is no wonder that we are so concerned, in western culture, to keep ourselves apart.

To transform our understanding of animal behaviour would mean relying less on the scientific accounts and, at the same time, it would mean making the scientific accounts less reductionist. Among other things this might require creating a firmer dialogue between scientists and others who work with animals: animal trainers, farmers, caretakers. These people often have a deep understanding of the animals with whom they work, but it is a knowledge gained from working with individuals, knowing their idiosyncrasies, rather than from studying animals in groups. This knowledge is typically denied by the formal practice of science. Yet those idiosyncrasies and the anecdotes of the wonderful things that animals have done are very much part of our cultural beliefs about animals. Moreover, they help to bring at least some kinds of animals *into* human society and culture, at least rhetorically.

Science, by contrast, typically refuses anthropomorphism and denies evidence unless it is the product of a specific procedure (an experiment) done by particular actors (scientists). Only such claims count as legitimate data in these contests at the human/animal boundary. Non-humans, in the scientific narratives, must be examplars of species, denied history and culture. They – but not humans – become cast as essences, as unchanging forms, evolutionary theory notwithstanding. As Benton (1993) has noted, Marx argued that only humans undergo development at the species level. The rest of nature, then, becomes incapable of such historical development.

While this may be true (and I would not wish to deny that humans have an immensely rich cultural history), I do wonder how we can *know* that other species' behaviour is so fixed. Indeed, the 'cultural' studies of chimp behaviour cited by McGrew (1992) suggest otherwise, at least for that species. Moreover, we cannot know for sure what animal behaviour was like in evolutionary history; behaviour rarely leaves artefacts, so there is little palaeontological evidence (the evolution of behaviour can only be inferred from taxonomic relationships and similarities).

Even in the timespan of human history, we cannot be sure. Yes, we have written records from earlier generations of naturalists, and these records do tend to accord with our perceptions of the behaviour of many animals today. But environments change, and animals must change with them. We also know that our perceptions are themselves cultural products; despite evolutionary theory, there remains a tendency to think in terms of animals as fixed essences, living in unchanging environments.

We also know that there is an intellectual lineage, such that observers of animals today are using perceptions that are themselves shaped by the beliefs of earlier observers. Would, then, we be able to see development at the level of a species, or cultural change? Or, if our views differed from those of our predecessors, would we simply assume them to be wrong, that science has progressed and we now know the truth? After all, part of our cultural belief *is* that non-humans are somehow fixed into nature. While biologists allow that some species might show 'cultural' change (variations in regional dialects of songbirds is one example), the prevailing image in biology is one of species-typical behaviour, relatively fixed in time.

Yet this seems to deny animals agency; to deny them any ability to transform their own environments except in ways that are themselves restricted by an equally fixed notion of their environmental niches (Master 1995). We need to emphasize that other species are active agents in the world, and that animal activities must have considerable effects on environments. Both species and niches must adapt to change together. Some biologists undoubtedly recognize this, but many do not, and the prevailing narrative is one that writes a story of animal essences.

Note

1 Small babies, however, presumably have similar nervous systems to those of an adult; yet, until recently, tiny babies were not routinely given anaesthetics, in the belief that they could not feel pain in the same way as an adult. Clearly, nervous system structure is not enough. I am grateful to Anne Fausto-Sterling for reminding me of this point.

Part IV

EMBODIED BEINGS

8

ANIMALS AND BIOLOGICAL DETERMINISM

In Part IV, I want to move from what science has had to say about non-human animals, and move on to look at the human/animal dichotomy in relation to feminist thinking about science. There are three ways in which I want to explore these intersections; the first area I look at in this chapter is the assumptions that we make in developing critiques of biological determinism. These critiques have, among other things, located non-human species in the realm of 'pure' biology, of mindless bodies. Chapter 9 takes up that theme of animal bodies, thinking about it in relation to ideas about the human body. Finally, Chapter 10 focuses more on asking questions – about science, about feminist thinking about science, and about animal rights.

Part of the background to feminist theorizing is the ways in which the social sciences construct ideas about gender, or about humanity. In general, feminists have concurred with much of sociology in being critical of ideas that imply biological determinism. Humans are some-how 'above' mere biology, being located firmly in culture; the realm of the biological – the body, the rest of nature – can thus conveniently be ignored.

Biological determinism

> the grand cause of hysteria . . . that which melts the women of England into powerless babes . . . and shatters their intellect . . . and, by destroying reason, level them with those that chew the cud – this grand traitor and foe to humanity is Polite Education.
>
> (Johnson 1850)

Anti-feminist arguments frequently seem to reduce us to the level of mind-lessness; in this case, the image invoked is of a bovine mindlessness, chewing the cud. In the middle of the nineteenth century, women's biology was held to be limiting; not only would it get in the way of our education, but education itself would damage our biology (Russett 1989) – thus reducing us to the status of animals. Women's intelligence, like animals, has typically been compared to men's – and found wanting.

Feminists have frequently had to confront arguments rooted in biol-ogy. Almost invariably, such arguments seek to justify the status quo. We are all too familiar with the force of claims stating that behavioural differences between women and men are rooted in genetic and/or hormonal differences. As a result, gender becomes immutable, fixed by biology. Typically, parallels are drawn with other animals (usually, but not always, other mammals); for example, the effects of hormones on rats' brains might serve to defend beliefs in fundamental differences between the brains of men and women (see Moir and Jessel 1989, for a recent example). The most common reactions to such arguments among feminists are to point to the inadequacies of the underlying science, and to emphasize the extent to which gender is socially constructed (e.g. Birke and Vines 1987; Hubbard 1990).

Politically, this move was important. All too often, women had been told that they should not or could not do something because of (say) their raging hormones, or because they were genetically/evolutionarily preordained to prefer the kitchen sink (for some examples, see Birke 1986; Hubbard 1990). The diet of biological forces continues, however, and has even gathered momentum in relation to homosexuality.

Feminist avoidance and rejection of arguments grounded in 'biology' is, politically, in striking contrast to an emerging groundswell of political feelings in the gay community around the question of whether homo-sexuality has biological bases. In 1991, for example, Simon LeVay, a gay neuroscientist, published a paper in the prestigious journal *Science*, which claimed to show that there were differences in the size of a particular tiny part of the brain (one of the interstitial nuclei of the anterior hypothalamus, or INAH for short) between heterosexual and homosexual men. The INAH in gay men was smaller, LeVay claimed, closer to the size of that of women (or, as he put it in a later book, 'gay men simply don't have the brain cells to be attracted to women' (LeVay 1993: 121)). LeVay's paper came in for criticism by, among others, feminist biologists, for its assumptions and for the quality of its science.

Subsequently further articles appeared, claiming further differences in the brain between straight and gay men (Allen and Gorski 1992) and claiming that there were identifiable genes associated with male homo-sexuality (Hamer *et al.* 1993). Whatever the quality of the science (and it is worth remembering that scientific papers can nearly always be subject to critical scrutiny on methodological or rhetorical grounds), what is interesting is the extent to which the gay community seems to have picked up on the biological claims. This seems particularly to be the case in the USA, where religious fundamentalism is rife; the political significance of this is that if gayness is born into you, then you cannot be held morally responsible for it – you cannot thus sin.

The development of these beliefs within the gay community is, perhaps, not surprising, given the power of homophobia. But it should be a cause for concern, partly because it perpetuates the myth that sexuality (or gender) is *either* entirely biologically caused *or* it is the product of social conditioning – or even choice. Partly, too, we should be concerned because there remains a risk that such theories play into the hands of those who seek to 'cure' those of us who are lesbian or gay. The claim that a gene has been found has already led to public suggestions that fetuses carrying that gene could be aborted.[1] Whatever the political responses, debate remains mired in a dichotomy: nature versus nurture, genes versus conditioning. Whichever side is emphasized, there are fundamental problems. Diana Fuss notes, for example, how social constructionism can seem to fail to speak to people's experiences of identity, particularly in the gay community, as it seems to deny shared history through its stress on historical contingency (1989: 106). If everything comes down to social construction/choice, then we have little left in common. That emphasis also fails to address our lived experience as embodied individuals; it is not a social construction when our bodies hurt and bleed.

Indeed, in our rush from biological determinism, feminists have sometimes seemed to deny biology altogether. Several commentators have pointed to ways in which 'the body' is denied; here I am more concerned with the non-human world, but the gap in the theory is there. Humans, in our theorizing, tend to occupy a largely disembodied world of the mind and culture; animals are located in a world of 'nature', of biological determinants, of pure biology.

Animals and biological determinism?

Biologically determinist arguments about humans typically rely on comparisons with other animals. This may be overt; it may be hidden away in the scientist's mind, but data from other mammals will be lurking somewhere. For example, although LeVay based his paper on data from human brains (41 to be precise), the hypothesis that there *might* be a biological correlate derived in part from studies using animals. Now it is certainly the case that these studies using animals can be subjected to considerable critique; what a rat does in laboratory studies bears little relationship to human sexual repertoires, for example. But the fundamental assumption that underpins this kind of work goes like

this: particular assumptions about human behaviour are made and pro-
jected onto animals – for instance, that what constitutes homosexuality
is a tendency to behave like one of the opposite sex. Then these findings
from animals can be used to 'show' that homosexuality is biological.
Then we can speculate that it is similarly caused in humans – and seek
a 'cure', perhaps testing that on animals first to show that it works.

All of this relies on assumptions that animals such as rats are an
appropriate 'model' for human sexuality. Yet those who argue using
biology seem mystified by the strength of opposition, as do those who
oppose biological determinism. Part of the problem, as with any controversy, is that the protagonists talk past each other. But another part
of the problem here, it seems to me, is the uncertainty (and thus implicit
disagreement) about what is meant by the terms. The most obvious
source of disagreement is over what 'sexuality' means. To those arguing
biological determinism, it seems to mean little more than penis-into-
vagina. To critics, human sexuality acquires meanings *in* human culture,
and can thus incorporate a wide range of behaviours and responses.
Perhaps it does to non-humans too, but biologists' emphasis on reproduction has ensured that animal sex is seen solely in that light.

Not only is sexuality a contested concept in these arguments; so too
are terms such as 'biological', 'animal', or even 'human'. For example,
in a discussion of what is meant by 'human' (by contrast to animal),
Cora Diamond (1992) notes the slippage in the way that 'human being'
may mean something to do with persons; it may have an existential
tone (as in human *being*), or it may be used in the 'biological' sense to
represent what is specific to our species. 'Humanity', moreover, means
more than just a species; it also represents a quality (we might speak,
for example, of the 'humanity' of decisions to help refugees). For various
reasons, people are now asking more questions about what that quality
represents, and so are beginning to challenge the human/animal boundary
more succinctly.

The term 'animal', as we have seen throughout this book, can mean
many different things. Opposed to humans, it tends to mean everything
we think we are not, or whatever we wish to transcend – the beast
within, for example. Colloquially, it can be used synonymously with
'mammal', as in the oft-used (but biologically strange) phrase 'animals
and birds'. To biologists, it means any organism that is classified as
belonging to the animal kingdom – thus a wide range of creatures, from
sponges and seaslugs to eels and monkeys.

'Biology' is also problematic. Obviously, there is one sense in which
it means a particular discipline, part of the natural sciences. 'Biology' in
this sense connotes the study of living organisms and their processes,
but it can also be used synonymously with those processes, as in 'human
biology'. It is this that feminist theorizing tends to avoid, although it is

not always clear just what it is about our 'biological processes' that is to be avoided.

The meaning of 'the social', of sociality, differs too, in these different discourses, particularly when animal sociality is invoked. In scientific accounts of animal behaviour, 'social behaviour' is held to be what a given species of animal does in groups – communicating, sexual behaviour, aggressive encounters, and so forth. But the 'social' of sociological and political theorists conveys much more than that; it is defined as not including animals. Rather, it is about human culture(s), politics and · ideology, and about the conscious design of these (Sydie 1987).

This is the crux of our belief in our ability as a species to 'transcend' our proximity to other animals. It is this that is uniquely human, we are led to believe, and which separates us from the rest of nature. Or separates some of us; as feminists have been quick to point out, it is not 'humans' who, through their involvement with the social/political domain, are more social than natural. It is men of particular backgrounds – typically white, western and middle-class. Beverley Thiele has pointed out, for instance, that,

> Male nature is independent, active and truly human while female nature, conveniently for the status quo, fits her only for a narrow domestic role . . . his nature is such that he may transcend his animality by escaping into the human political realm of civil society, but only because women, trapped by their biology into remaining in the private sphere, oversee all the animal-like functions of mastication, defecation and copulation.
>
> (1986: 37)

Small wonder, then, that feminist theory has so often entailed denying proximity to animals/nature, as we seek to escape that world so dominated by 'animal-like functions' (while, paradoxically, many of us share our lives with non-human animals). Despite the sexism of which Thiele complains, the

> notion of a socially constructed subject . . . is . . . absolutely central to feminist theory . . . To view human being [*sic*] as a social product devoid of determining universal characteristics is to view its possibilities as open-ended. This is not to say that human being is not constrained by historical context or by rudimentary biological facts but rather that these factors set the outer parameters of possibility only.
>
> (Gatens, 1986: 28)

Here, then, is the reason for the commitment to social construction within feminist/sociological theorizing; it implies open-ended possibility, by contrast to the constraints set by 'rudimentary biological facts' to which, implicitly, *non*-human organisms are subject.

That only humans have true sociality is an assumption threading many political perspectives, from left to right. Thus the right-wing philosopher Roger Scruton argues that animals cannot be 'really' social, suggesting for this reason that the use of the word social in descriptions of animal behaviour is inappropriate:

> Even the description of the herding animals as 'social', in so far as it implies a certain conception of 'self' and 'other' through which relations are mediated, is a false designation of their behaviour. That the organisation of a clan of gorillas *looks* social is obvious; but where are the laws and institutions, where the adjudications, where the disposition of rights, privileges and duties, which make up the social consciousness of man? Without rationality such things can never come into being.
>
> (Scruton 1986: 59; emphasis in original)[2]

Those who become defined as being outside rationality cannot, it would seem, be truly social. That is defined as any being who is not human.

The assumption in western thought is not only that humans are different from 'other animals', but that what we have/are is superior (as in the above use of rationality). Now, I would not wish to argue that human cultures have not generated qualities that are different from any other species. Perhaps we have produced many wonderful things (not to mention such appalling things as the capability for global destruction), but this kind of formulation, which sets up particular characteristics as defining humanity's specialness, is asking for trouble.

The biology versus social construction opposition requires that we firmly separate ourselves from other species. Biology is what non-humans have, while we disappear off into realms of virtual (social) reality. That separation runs strongly through the social sciences, and has influenced feminist thought. Thus Chris Shilling (1993) outlines the importance of recent sociological thinking about the (human) body, and how we should not relegate it to the realms of the presocial or biological. Yet he retains the biology/culture divide when it comes to animals versus humans, suggesting, for instance, that

> [a]nimals enter the world with highly specialized and firmly directed drives . . . [they] have a species-specific world whose territories and dangers are mapped out from the very beginning of their lives. The bodies of animals are programmed to exist and survive within their environment.
>
> (Shilling 1993: 101)

This picture of programmed automata is not a picture that accords with my understanding of many animals.

There are two particular consequences of the animals/nature versus humans/sociality dichotomies that I want now to examine. The first concerns their roots in attempts to define what is human by setting up 'others' who are not admitted into the human sphere; the second focuses on the problems of determinism and reductionism for how we see *non*-humans – the kind of reductionism that continues to suppose that other animals are 'programmed' firmly by the dictates of biology.

Animals and others

Separating ourselves from less worthy 'others' is, as we have seen, a common trait. Feminists have been critical of this separation, not least because women have so often been other to men. 'Othering' is a trait that links to concepts of global domination, in its reliance on defining what is human against those who are conquered (of whatever species). Rey Chow (1989: 158), for example, points out how the 'efforts to delineate the "human", be it in terms of language, reflection, or subjectivity, took place against a background of looming non-European "others"' who remain part of nature (see also Halpin 1989).

The humanist traditions in western thought have tinkered with the boundaries of domination, with what is included or excluded. At various points, different classes of people have shifted from inferior status to claim inclusion in humankind – non-Europeans or women, for example – but none of this shifts the underlying structures and assumptions; all it does is to widen membership of the dominating group (Plumwood 1991). Nature remains firmly outside.

In rejecting biological determinism (however right our arguments may be), while not questioning what that means for and how it depends on particular ideas of animals, we are setting ourselves up as different from/ superior to/apart from them; the assumption that our behaviour has very little to do with biology is aligned with one saying that theirs has everything to do with their biology. What we are saying, in effect, is that *we are not like animals*. Why do we want to say this?

Animals in western culture are 'other', objects of scientific enquiry (Halpin 1989). We have defined ourselves in opposition to a generality of 'animals', irrespective of the qualities of individual species. It seems paradoxical that at a time when much feminist theory is moving beyond simple dualisms of gender (putting great emphasis on differences between women, say), it should do so by implicitly building its analyses on another simple dichotomy – humans versus 'other animals'. A more consistent approach, indeed, might be to extend the emphases on plurality and difference, and to begin to deconstruct the (putative) boundary between us and other species.

Determined animals?

What we typically do in critiques of biological determinism is, rightly, to object to the way in which simplistic ideas about 'the biological base' are used to justify women's subordination. But what we fail to do is to question sufficiently the premises; we thus do not ask questions about the meaning of the biological body, nor do we ask questions about the behaviour of other animals. What happens, implicitly, in attacks on biological determinism is that the role of biology in bodily functions is not challenged; only its alleged role in our behaviour and capabilities is at fault. So, few would deny that ovaries produce hormones that are involved in the events of menstruation or menopause.

My problem with that is that the body and its internal functions so often seems to become primary and presocial (Shilling 1993).[3] The critiques of biological determinism, moreover, tend to rest on two dichotomies: mind/body and human/animal. To talk of biological substrates only becomes a problem in these critiques when it applies to humans, and minds. Thus in critiques of biological determinism, the behaviour of animals is rarely seen as problematic; we might, for instance, object to the notion that human gender differences are determined by hormonal effects on the brain while accepting a similar notion from laboratory studies of other species. So, from this perspective, human behaviour is fundamentally different from (discontinuous with) our biology, and human experience is radically different from that of other animals.

One difficulty with assuming human behaviour and abilities to be different from that of other kinds of creatures is that it leaves the lives of all *other* organisms firmly within the jurisdiction of 'biology'. Books on the biology of rats, for instance, are likely to include something about their behaviour and social organization, as well as how ovaries work. This allocation has two problematic effects. First, locating animal, but not human, behaviour within the realms of biology can distort our perception of the behaviour of animals. Second, because the predominant mode of explanation within biology still tends towards reductionism, it is all too easy to see non-human behaviour and social structure as caused by an underlying 'biology' – genes, hormones, or whatever. (This is, however, more obvious in popular accounts. Some areas of biological research have moved away from such simplistic models, towards a recognition that animals do not fit readily into the deterministic picture either. My purpose here is to draw attention to the fact that, if the behaviour of some species is called 'biological', then that behaviour is more easily attributed to 'biological' underlying causes – both in that species and then in humans.)

Biological explanations do tend towards reductionism, a point underscored by feminists often enough. But if reductionism impoverishes our understanding of human behaviour, then surely it must also do so with regard to animal behaviour. I have argued elsewhere (Birke 1986; 1989) how different assumptions underpin questions about gender difference in humans and 'sex differences' (a telling phrase, for it implies fixity) in animals. In animal studies, scientists may note a sex difference in some mode of behaviour; usually this means a difference in the frequency of some kind of behaviour between populations of males and populations of females. Occasionally the frequency in one population is almost zero; more often, the populations differ merely in the relative frequency of that behaviour. For example, male rhesus monkeys may initiate social play more often, on average, than females, but that does not mean that females never engage in social play; nor does it mean that the animal's sex is the only thing that matters – bodily size may be important. Females tend to be smaller among rhesus monkeys. I would certainly hesitate before leaping in play onto a larger male – I imagine that a rhesus monkey would, too!

Now none of these differences are absolute, and they are statements about populations. The next step in reductionist logic is to extrapolate backwards within the developmental histories of individuals, and to look for antecedent causes within those individuals. You start, in other words, with an average abstracted from a population (male rats are more likely to attack strangers than are females, say), and then draw conclusions about the life history of an individual within that population. Because each sex produces different levels of certain sex hormones, those are a popular candidate. So, you might ask whether exposure to particular hormones in infancy makes a male more likely to attack as an adult. Once the hormone has been implicated in laboratory rats, then it is likely that someone, somewhere, jumps to conclusions about people – as indeed has been the case with the tales of hormonal causes of homosexuality.

But why assume that hormones within individuals are the direct and sole cause of any difference between populations? Hormone levels fluctuate in response to social or environmental influences in other mammals just as they do in people. If we can point to variation as a result of social influences in ourselves, then why not in other animals? Many of those who study animal behaviour do so, but not all, and the popular image of animal societies as machinery writ large persists.

We do not know enough about how sex/gender differences emerge in animals out of their previous experience of social interactions. I would certainly accept that hormones are involved in behavioural development in some animals, yet that is not to say that the role is determining; within a large litter anything that happens hormonally to one pup before or after birth has repercussions on the others. It is not an individual that the hormone affects, but a social system. For example, mother rats respond to each pup on the basis of its scent, which in turn is affected by its hormones. So if scientists have interfered with a rat pup's hormones, how would they know whether any subsequent change in its behaviour were due to the hormone, to the mother's behaviour, or to some complex interaction of both? (see Moore 1984; Birke 1989).

From this perspective we can allow behaviour, even in other animals, to become more than the product of internal 'biology'. We can begin to emphasize the richness of an animal's individual development, and its engagement with its own environment – as well as considering how that engagement could in turn influence the 'biology' itself (hormone levels, say). Acquiring gendered behaviour can then become seen as a more complex and negotiated process – even for non-human animals. We would also then, significantly, be breaking down the logic that locates animal behaviour in a reductionist biology, while retaining human (gender) behaviour in the social domain.

Part of my reason for questioning our assumptions about other animals is, of course, that I have sufficient respect for non-humans to find the reductionist accounts ridiculous. Like rhesus monkeys, male rats tend to engage more often in social play than females. But I have watched playful young rats often enough to feel that this is not simply due to having hormones of a particular type. The differences seem to *emerge* from social interactions within the group, between mother and infants, between one littermate and another. If I, the scientist, focused on female/male differences at the time, then that is because my scientific training had taught me to think in such ways; in doing so, I failed for a long time to see the complexity that was in front of me.

Questioning assumptions about animals and biology is also partly to do with wanting to question the human/animal (or human/nature) dichotomy and its grounding in assumptions about human domination in all its forms. Partly, I seek to do so out of respect for other animals for their own sakes. Other animals are not simply a package of territoriality or other 'drives', but complex, decision-making creatures engaging with their environment. Another reason is that, if we continue to leave the deterministic assumptions about non-humans unchallenged, then biologically determinist arguments about us can easily flourish. For looking to other animals to support such claims lies at its heart. Simplistic accounts of animal societies combine with various normative assumptions derived from our own society to yield a powerful picture of coy submissive females and dominant, pioneering males. We ignore the other side of that picture at our peril. Treating other species as merely puppets of their genes feeds straight back into the biological determinism around human behaviour that feminists so much despise.

Rewriting subjects

So where does all this leave the 'specialness' of humans, and the assumptions made about that specialness by feminist criticisms of biological determinism? We can certainly argue for human specialness; there
may well be features of our social life and culture that no other species
at present possesses (just as we may lack certain features that other
species possess). But we should not, I have argued, do so by creating an
animal mythology. Animals are not automata, subject only to the dictates of biological laws; many species lead complex and rich social lives
that cannot be collapsed onto simplistic ideas of biological substrates.
Nor are we merely social constructs, beings without a biological body.

If we are effectively to counter biological determinism, we have to
examine the various assumptions we make about our relationship to the
natural world. We cannot simply escape them by dumping some bits,
but not others, into a rubbish bag called 'biology'. To acknowledge a role
for our biology – as bodies, or our part in the natural world – has sometimes seemed a dangerous move for feminists. And so it is, if we leave
unchallenged the assumptions about what constitutes the biological or
about what animals are.

We must, of course, continue to challenge crudely biologically determinist arguments. But we must also remember that determinist arguments do not work even for the 'other animals' that form the implicit
counterpoint to our rejection of biological arguments about women. Our
understanding of the natural world has, as a result, been impoverished.
We must, therefore, insist on seeing (any) other species or individual as
a product of complex life histories, in which all kinds of factors interact
and play their part. We should strenuously deny that non-humans are
'simply' the product of bits of biology, just as much as we do about
women. One of the main problems for feminists about biologically
determinist explanations of our behaviour is that these explanations
deny us agency, as Helen Longino (1989) has pointed out. They deny
agency and subjectivity to other animals, too.

The descriptions of how animals behave that we find in scientific
accounts undoubtedly yield predictions (not to mention natural history
television programmes). They are also useful as a basis for decisions
about how to evaluate potential animal suffering (Dawkins 1980). But
there are limits to their usefulness. The scientific stories go nowhere
near explaining how, for example, some species of birds find their way
across thousands of miles – demonstrations that they respond to the sun's
position, or to stellar configurations are, I suspect, only a small fragment
of the story. They go nowhere near explaining the abilities of sheepdogs
to anticipate the movements of flocks of sheep. They go nowhere explaining how both I and the horses I ride can learn a complex tactile

form of communication (a language?) that enables us to do a range of athletic actions in harmony.

These last two examples are ones of non-human animals who are already, in some senses, *in* human society, and who accomplish these feats in relationship to a human being. That is not to diminish the activities; it merely points to the value that we humans place on animal skills when those skills benefit us directly. What these examples do illustrate, however, is the potential richness of animal experience when it is not constrained by the kind of environmental impoverishment that pertains in laboratories. In those conditions, we get a very different picture of 'the biology of' non-human animals, defined primarily by their *in*abilities rather than their potential abilities.

In her book, *Humans and Other Animals*, Barbara Noske suggests that, rather than adopting the objectifying stance of laboratory science (as feminists implicitly do towards the animal world), we should seek to define an 'anthropology of animals', which allows them to be active subjects (Noske 1989). Biological determinism relies on picturing animals as crudely driven by biological imperatives; at its worst, these imperatives are simply attributed to humans too. Neither is true. It is not enough for us to rebut only the human version of the story, for that tale relies heavily on an animal fable.

Notes

1 The claim met with considerable outrage, not only for what it said, but also because it was made by London's Chief Rabbi, who seemed to have forgotten that many homosexuals also met their deaths in Nazi gas chambers.
2 It is these institutions which create the conditions, argues Scruton, for a truly human sexual desire (animals, he says, have only sexual urges). This is the basis for his arguments in favour of conventional morality (in, for example, marriage).
3 This is not to deny that, in some feminist theorizing, the body has been understood as socially contingent. We can understand 'bodies' only through our social and cultural perceptions of them (see Hekman 1990). But my reading of this kind of theorizing is that it seems to focus on the outer surfaces of the body and the meanings inscribed thereon. This (to me) still leaves the inner processes of bodies (which are the concern of feminist critiques of biological determinism) as relatively untheorized.

9

DENYING THE (ANIMAL) BODY

Whale: I talk about my body because most humans are so ignorant. You don't know much of what you should about the varieties of motion and stillness, or pain, or breathing, or dreaming, or any of the other essential processes of the body . . .

Woman: It is the language of the body. I know she understood me. I didn't speak with my mouth. It was more reaction: my teeth shook, my body arched in the water, my face changed shape. None of this is silent. She heard what I meant.

(Stinson 1990)

Several feminist writers have identified within western traditions a deep fear and hatred – somatophobia – of the body (Spelman 1982); we can thus become ignorant of our bodies, as the whale implies. What is most despised in these cultural traditions are precisely those functions that we share with animals, the 'animality' of our bodies as they sweat, bleed or defecate.

Both animal bodies and human bodies have in common the way that they become, in some scientific discourse, incidental to other ends. In the rhetoric of physiology, with its various systems – the cardiovascular system, the central nervous system, and so on – the wholeness of the body and its relationship to its own environment seem to slide out of focus. Similarly, in modern genetics and evolutionary theory, what seems to matter is the gene pool, as we have seen; it is genes that reproduce themselves into future generations, not bodies. And it is 'the gene pool' that we might try to protect if we work for endangered species. It is not the integrity of individual bodies that matters here.

Whatever the 'essential processes' of the body are, they tend to be seen in western culture as the ineradicable *base* of who we are – whether we be whales or women. Humans, however, have something added to that base, an essence of humanness. One of the problems for thinking about notions of biological determinism in relation to feminist theory, as we have seen, is that we cannot simply ignore the rest of nature or our bodies in our zeal to describe social constructions. 'The body' is also, as feminists have often pointed out, culturally coded; bodies are not just bodies, but male or female ones, with all the meanings that entails. Bodily parts, too, are coded; having large well-defined muscles, for instance, is associated with masculinity. Bodies stand as powerful metaphors, representing boundaries of class, gender and race – or, in the metaphor of the 'body politic', standing for the (masculine) state itself (Gatens 1991).

What I want to do here is to use these feminist explorations as a framework within which to explore some of the meanings of animal bodies. In science, for example, to talk of something like 'the vertebrate body' is to speak of particular meanings of generalized body plans; vertebrate bodies contain particular features (such as a backbone) built around definable spaces (the thoracic cavity for example) and shapes (they are typically symmetrical). These body plans are represented diagrammatically, as sections through the body, to produce the highly abstract, almost 'architectural' representations of the scientific textbook rather than artistic renderings of what the observer sees in nature (Laqueur 1990). The neat line drawings of stylized 'body plans' of vertebrates represent a distillation, a profound simplification, of the knowledge we have of how bodies are put together – a kind of artistic rendering of scientific reductionism.

Meanings are also created in other spheres; animal bodies have cultural significance outside of the physiology labs, as we have seen. They also have meanings within the labs; the images invoked by reading in the history of physiology are images not only of heroism (the scientist-hero struggling with the vagaries of nature to understand how, say, the heart works), but are also gruesome. These are the images of animal bodies spread out on the dissection table, pinned down and displayed to the gaze. Before the advent of anaesthesia, these were also images of living animal bodies thus displayed, of howling dogs and tortured cats (for examples of paintings in this vein, see Schupback 1987).

I begin this exploration, then, with a brief look at the history of the idea of 'the beast-machine' which has historically dominated physiological research. The imagery of the body as a machine (but, if human, a machine with a soul) persists today, in scientific writing and iconography. From there, I turn to ways in which ideas about animal bodies and human bodies both differ and are similar, before returning to the broader themes of the animal/human dichotomy and its relationship to politics.

The beast-machine

But Since I was once persuaded that Beasts were destitute both of
Knowledge and Sense, scarce a Dog in all the Town, wherein I was,
could escape me, for the making of Anatomical Dissections, wherein
I myself was Operator, without the least inkling of Compassion or
Remorse.
(Father Gabriel Daniel 1690; quoted in Maehle and Trohler
1987: 28)

The notion that animals are mere automata, 'beast-machines', is most
associated with the name of the French naturalist-philosopher, René
Descartes. To him, only humans had a soul or mind, and therefore the
ability to think. Animals are mere materiality, simply mechanical bodies.
Human bodies, too, are mechanism; it is just that we have these mys-
terious minds thrown in to render us unique. It is the mind/soul,
moreover, in Cartesian doctrine which allows humans – and only humans – to feel pain. According to Descartes, animals may behave as if
they feel pain, but without experiencing its mental sensation. This doctrine appears to have been used as justification for the use of animals in
physiological experiments, according to several accounts from the late
seventeenth and early eighteenth centuries (see Radner and Radner
1989: 98; Maehle and Trohler 1987). It was just this attitude towards
the use of animals that Father Daniel satirized in his account.

There were certainly many dissident voices who questioned the idea
that 'beast-machines' were insensible to pain. But the practice of physiology was easily justified by recourse to such concepts. Elliott (1987)
notes the uncompromising stance taken by the French physiologists of
the nineteenth century. In developing a specifically *experimental* physiology, founded on the use of living animals, these physiologists could
'receive the unquestioning support of the élite of scientists at the Academie
des Sciences' (75). This, combined with the Cartesian concept of mechanism, ensured that animals used in research would be treated as a
largely insensible part of the apparatus of science, as mere machines.
The basic notion that they were insensible to real pain was modified
somewhat by the late nineteenth century (partly in response to anti-
vivisectionist feeling; Rupke 1987), once anaesthetics were available and
the law in Britain required their use for surgically invasive procedures.

Alongside the concern of physiologists to look inside the animal body
was a growing demand from doctors for human bodies to dissect. Here,
too, the power of the medical profession is an important part of the
story. Demand for corpses exceeded supply by the end of the eighteenth
century, leading to a rising tide of grave-robbing and even murders. As
a result, legislation was eventually enacted in Britain, in 1832, which

permitted doctors to use the bodies of those who had died 'unclaimed' in the much-feared workhouses – leading, not surprisingly, to public unrest (Richardson 1989). Not only did people feel that dissection was worse than death itself, but they were well aware of ways in which richer people (including doctors themselves) could avoid such a fate by paying for tombs. In that sense, the physical bodies of the poor came to represent in death their struggles against a powerful establishment. Like Richardson, I have met people while I was growing up in London who still save towards a 'proper' funeral, fearing the consequences of a pauper's burial – a cultural echo of earlier generations' fears of being dismembered after death. In the light of that history, it is not perhaps surprising that one part of the opposition to animal vivisection a century ago came from working-class people in London who feared that 'it would be us next' (Lansbury 1985).

These developments in the practice of physiology and anatomy were taking place against a background of increasing scientific interest in classification and a scientific curiosity in the unusual. Richardson notes the sad story of the 'Irish Giant', a man of over seven feet tall, who had requested burial at sea. But large sums of money ensured that his body was taken from the coffin; his skeleton still stands as part of the Hunterian collection of the Royal College of Surgeons in London.

Body plans of different animal kinds were also a consequence of this fervent interest in cataloguing nature. What was changing during this period was the ways that animal and human bodies were perceived. Where once superficial physical difference had been emphasized, scientific beliefs in underlying similarity of processes became the norm. Doctors could thus experiment on living animals alongside their use of human corpses in their studies of anatomical structures. The scientific world view we have inherited today is thus a system emphasizing similarity of underlying processes rather than of the external appearances of bodies (Tester 1991) – a representational dismemberment of bodies into their constituent systems. All vertebrates have broadly similar features of respiration for example. What the physiologists of the nineteenth century produced, then, were stories of functional systems, from which bodies were built just as we might build a machine from component working parts.

The notion of bodies as machines persists today. Physiology textbooks contain references to analogies between engines and how bodies work, or between bodies and factories; the practice of medicine, too, seems to see us in terms of the parts. Thus a textbook from my undergraduate days (tellingly called *Living Control Systems*) began with the statement that

Animals and plants are chemical factories . . . Animals, in addition, are provided with engines which enable them to move about: the factory can move, when necessary, to its source of raw material, or escape from rivals who seek to devour it.

(Bayliss 1966: 1)

The factory metaphor pervades many descriptions of the body and its functions, as Emily Martin has stressed in her account of women's perceptions of their bodily functions (Martin 1989). The factory metaphor emphasizes efficiency and control – but this can break down and become inefficient. Gynaecological accounts of women's bodies, she notes, portray menstruation in terms of the failure of the body to become pregnant; menstrual bleeding and the menopause are described in textbooks in terms of degeneration, failure and decline (she also notes the contrast with the heroic journey of sperm, which is described typically as 'amazing' or 'remarkable'; also see the Biology and Gender Study Group 1989).

Apart from the assumptions about gender, such images and rhetoric of bodily function serve to perpetuate a notion of non-human animals as being little more than mechanisms. On one hand, this kind of imagery stresses the similarity between humans and other animals; physiological principles are said to apply to all living things, so that the human body works in some similar ways to (say) that of a flatworm. Both, for example, rely on exchange of gases in respiration.

On the other hand, difference between human and animal can be acknowledged within the metaphor, precisely because the factory narrative is hierarchical. What differs is what ultimately controls the hierarchy; however much scientists may wish to avoid such notions as mind, it is precisely human 'minds' that can be invoked to save that much-prized human superiority as controller-in-chief of the factory complex. And if we wish to remain mechanistic, we can always claim that human brains (the machinery of mind) are better than anyone else's at such industrial management.

Not only are bodies factories in some texts, they are also technological communications systems in others. Nerves and hormones help to execute communications within; the immune system becomes a 'fluid and dispersed command-control-intelligence network' (Haraway 1991b: 211). The body, in this emerging set of discourses,

is conceived as a strategic system, highly militarized in key arenas of imagery and practice . . . The privileged pathology affecting all kinds of components in this universe is stress – communications breakdown. In the body stress is theorized to operate by 'depressing' the immune system. Bodies have become cyborgs – cybernetic organisms – compounds of hybrid techno-organic embodiment and textuality.

(Haraway 1991b: 211, 212)

These metaphors of command and control help to perpetuate an image of body as mechanism, as machine – at once mindless and capable of strategic planning. The experimental methodology, moreover, has helped to generate a language of physiology that denies the sentience of non-human animals. The animal body, in these narratives, is composed of systems that, for example, maintain fluid balance, or serve internal communication. Even the brain becomes reduced to sets of structures controlling different functions. These are the living control systems of student textbooks, the product of a science that has assiduously denied mind, intentionality or purpose to living organisms. There is no 'naturalistic animal' here, only an 'animal' stripped of its literal and metaphoric existence.

The language used, in scientific reports, to describe experimental results is part of this process; as we have seen, it is a language that is abstracted, denying agency even to the scientist performing the techniques. Statements such as 'heart rate was measured on isolated, perfused heart preparations' dislocate both the reader of the text and the organisms involved (human or animal). The reader has to reinsert the meaning: the scientists killed the animals, and then removed the heart and measured the rate of its beating. The coldly abstract style of scientific writing codes the animal's body as merely the box in which the heart was originally found.

Physiological knowledge is structured largely through looking at the bodies of other kinds of animals – at a distance, and usually dismembered. Even medical training assumes the dispassionate stance when human corpses are dissected. There are few situations where we can experience what it would mean to look, not into 'the' body as representative, but into *my* body. Women might look at ultrasound images of 'their' fetus, but those images focus precisely on the separate existence of the fetus – indeed, the images themselves seem to be taken to *be* reality (and called 'photographs of your baby'), playing down the role of the body of the pregnant woman herself (Petchesky 1987). Indeed, it is because of the scarcity of ways of looking into our own bodies that the focus on vaginal self-examination in women's health groups in the 1970s became a radical act; they enabled us to see inside *our* bodies, our own flesh.

The cultural coding of bodies

Both animal bodies and human bodies are culturally coded in many ways. As we saw in earlier chapters, particular kinds of animals have come to represent particular sets of values; thus various breeds of dogs carry different meanings within western culture. What I want to explore here however is not so much the general meanings of 'animals' in our culture, but to look at some of the ways in which meanings and representations of animal and human *bodies* overlap. The human body is undoubtedly replete with symbolism, but so, too, are at least some kinds of animal bodies.

One such area of overlap concerns the malleability of the body. Although we may tend to see the body as fixed, we also recognize that it can be changed, within limits. This is the basis of cosmetic surgery or bodybuilding, for example. Cosmetic surgery relies on beliefs in the perfectability of the body towards a culturally prescribed standard. Humans may impose this on themselves (if they are affluent enough to pay for the surgery involved); they may also impose it on companion animals. 'Breed standards' for particular kinds of dogs, for instance, have traditionally involved tail-docking, a mutilation that can also be thought of as a form of cosmetic surgery (admittedly according to human standards of 'perfection').

In part, the pursuit of bodybuilding, like the pursuit of the slender body (Bordo 1990), is also about seeking a particular image, about seeking perfection of, or in, that body. Perfection implies an underside, however, a kind of body that is less than perfect. Whatever the perfect body is in our culture, it is undeniably slender, fit – and above all, able-bodied. Marsha Saxton, writing about the implications of prenatal diagnosis of genetic disease for people with disabilities, notes how our culture encourages us all to 'achieve rigid standards of appearance. Such standards are particularly harsh on disabled women whose appearance or body function may be further from "acceptable"' (Saxton 1984: 306).

Another counterpoint to the perfect body in our cultural imagination is the 'grotesque' body, which stands in contrast to the classical masculine body, as part of civilized culture (Morgan 1993). The grotesque body in this cultural contrast is, suggests Morgan, much closer to nature. Historically, it has been associated with peasants, it is more feminized. It is a figure that emerges as the symbolic Wild Man of European mythology, or associated with monsters.

The grotesque body as a monster is a powerful icon in western imagery; typically, it is part human/part animal. In medieval writing, this creature may be the unhappy result of copulation between human and nonhuman, or of practices that were 'contrary to nature' (such as sodomy or other non-procreative sex). Apart from the powerfully normative values these monster hybrids embody, they also illustrate a cultural 'horror of monsters', and a desire (underlined by Christian theology) to separate ourselves from the beasts (Davidson 1991). Whatever else the grotesque body is in western mythologies, it is partly an *animal* body, which our culture has needed to separate from idealized images of, for example, Michelangelo's *David*.

The recurrence of this figure as a theme in western cultural history reflects a deep-seated anxiety about defending the boundaries between humans and the rest of animal-kind (and, significantly, between the idealized male body and the bodies of women). In more recent times, anxiety about transgressing the human/animal boundary centres upon the possibility of the creation of monsters through scientific intervention – through DNA hybridization techniques, for example. A popular television drama in Britain played on this fear; entitled *Chimera*, it featured power-mad scientists creating an ape-human hybrid who eventually broke out of the laboratory to cause havoc. The film *Jurassic Park* plays on similar fears. Here, the monsters do not transgress the human–animal boundary so much as transgress the boundaries of nature; the film revolves around the question of whether scientists should have 'let sleeping dinosaurs lie'. Is it right to recreate these prehistoric monsters to terrify us all, to break the long sleep that nature imposed when the dinosaurs died out? Monsters remain a powerful cultural symbol of animals as beasts; these are grotesque and terrifying bodies out of control, a symbolic threat to the integrity of the human body *in* civilized culture.

'Perfecting' the body by dieting or in bodybuilding, by contrast, is about controlling it. Bodybuilders push themselves to limits, controlling their food intake and the work they impose on their muscles. Bodybuilding through athletic work carries another set of representations; athletic work builds muscles which carry powerful meaning. Well-developed muscles are culturally coded as masculine, so that women bodybuilders are acceptable only within certain limits of muscle definition and 'femininity' (Kuhn 1988; Mansfield and McGinn 1993). But it is not only human muscle that is so coded. The horse, that icon of power, is one animal whose athletic potential humans have developed, and there is now big business in scientific study of equine athletic performance (Snow *et al.* 1983). When horses stand as icons of power, they are muscled, hard and fit; they are also male. I have only to think of the various statues I have seen in different capital cities of Europe, displaying the conquering (male) hero, aboard a fighting horse of equally unambiguous gender. Not only is the animal invariably well equipped with muscle, it is also well endowed.

The sexed body

Both animal bodies (of whatever type) and human bodies share at least some meanings. They are both seen as systems of mechanism; they both constitute fixity. That, after all, is at the heart of distinctions between sex and gender; sex is seen to be the biological base, the fixed and unchanging core. So, if sex is the core for both animal and human bodies, then it makes sense to use animal models for studying 'deviant' human sexuality. I have long been disturbed by claims that scientists have created 'homosexual' rats by tinkering with their hormones. The animals in this case are not given the benefit of the doubt; perhaps, after all, some of their sexuality is expressed in multifarious ways? On the contrary, this is about perpetuating images of deviancy, of animals with aberrant hormones. 'Normal' male rats mount; 'normal' females stick bottoms in the air to allow a male to mount. All else is non-reproductive deviancy.

So in this world of certainty, of one thing or the other, alterations to such certainty must mean that the gender of the animals is compromised. A male that doesn't fit the stereotype cannot be a real male; therefore, the reasoning goes, it must be homosexual. Note that it is not that the animal indulges in a bit of homosexual behaviour (however that may be defined for a laboratory rat); what is at issue is that the animal itself, its *body*, becomes coded as 'homosexual'. Thus, scientists may refer to 'a' homosexual rat; the animal becomes coded, and thus its body. These are the (male) rats who permit themselves to be mounted, or the females who like best of all to mount. No matter that it is, respectively, males who will mount the 'deviant' males, nor that it may be females who will permit the mounting by the 'deviant' females. These other animals are not homosexual in this rhetoric.

Animal bodies, and what they do with them, are thus rigidly dichotomous. What matters in this story is sperm meets egg, boy meets girl; biology would have it no other way. The metaphor of animal bodies as reproduction machines is powerful; indeed, that is one reason why we might react to the notion of 'homosexual' rats as an absurdity. But this is as much a product of cultural ideas as it is about nature; science follows western cultural traditions that have eschewed (indeed, violently oppressed) any non-reproductive sexual expression.

The imagery of the 'homosexual' rat inevitably feeds into and reinforces similar imagery about humans. Medical journals in the 1950s and 1960s contained many articles referring to 'the lesbian body' (see Birke 1980). Researchers, collapsing lesbianism onto gender, assumed that the lesbian body must be masculinized; thus they sought evidence for broader shoulders or less regular menstruation among lesbians than among heterosexual women. 'The' lesbian body, in these medical myths, is as much a fiction as the homosexual rat of the laboratory researcher. But each, as a representation in scientific texts, owes much to the other – though I must add that (unlike some genres of fiction) neither speaks to any of my erotic, bodily or emotional experiences as a lesbian.

When we speak of 'the body' we thus invoke many meanings. To talk of 'the' human body is both to speak of specificity and of generality. It is specific, in the sense that the qualifying 'the' might differentiate this kind of bodily form from any kind of 'animal' bodies (of many different forms). It is less specific, in the sense that 'the body' as a shorthand glosses over its nature as a sexed body. Until recently, many books on anatomy typically featured male bodies, with female ones thrown in only to show how they deviate from the male norm (I have an *Atlas of Human Anatomy* on my shelves in which female bodies are rare indeed). But 'the' body is also generalized; it may be my body that I experience, but 'the' body connotes a general pattern – a structure composed of various systems.

Similarities and differences

Yet despite the similarities in the ways that we might think about human and non-human bodies, there are also differences. We share much with other mammals, but we are also the only species of mammal that habitually walks on two legs (with all the advantages, and problems, that result). There are many ways in which we could speak of differences, but here I want to focus on some differences in the way we might see animal versus human bodies.

One way of speaking about difference in meanings attached to animal or human bodies concerns the emphasis placed in some recent feminist writing about the (human) body as itself socially constructed. That is, we cannot talk or think about or even experience 'the body' *except* through its social representations. To take one example, Susan Hekman rejects the biological determinism/social construction opposition, suggesting instead that

> we can think in terms of biological sex as something we understand *through* social categories . . . Biological sex and socially constructed gender are not separate or opposed, but, rather, form an integral part of what we are as individuals.
>
> (1990: 142)

In this context, she cites Helene Cixous' argument that 'our bodies are themselves social constructions'.

My main concern about these claims is that they are fine – as long as your body is not that of an animal. However much 'animals' might become social categories, or enter human society, their *bodies* seem to be left beyond these arguments. 'The' animal body seems to remain firmly in the realm of biology: the unsocial machine. That the body in question is human is taken for granted in most of this writing, even when it is carrying out 'animal-like' functions such as defecating. Animal bodies are the hidden 'others', a counterpoint to feminist social reconstructions.

The language of social constructionism, with its emphasis on understanding difference and contingency, stands in stark contrast to the certainties, 'facts' and alleged similarities of scientific accounts. Take the example of menstruation. We can point to the ways in which, for example, the experiences of menstruation are socially constructed; thus women from different cultures may experience the menstrual cycle differently. But that phenomenological account leaves the discourses of science untouched; in its concern with the act of experiencing, what happens to the stories of hormones and bleeding? In Emily Martin's study of women's perceptions of their bodies, she noted how different women related to the scientific accounts of menstruation in different ways. Thus women from working-class backgrounds tended to use an experiential account, perceiving the onset of menstruation (menarche) in terms of a passage to womanhood. Middle-class women, by contrast, tended more explicitly to produce an account in terms of what they had learned of the medical model – the hormonal events, for example (Martin 1989). The experiences of bleeding, of discomfort or pain, become subordinated for these women to the certainties of scientific facts.

Another area that presents contradictions has to do with the health of the body and the politics of health and disease. 'The' body can be experienced as functioning well (health) or as not doing so (disease). Many feminists have, rightly, emphasized the various ways in which our lived experiences (of, for example, poverty) profoundly affect our bodily experiences of ill-health. Yet at the same time, we must rely on 'factual' information about disease and its treatments derived from highly constrained studies of animals and their bodies, in which the animal body becomes cast as a 'model system' – a walking set of heart and lungs on which to test drugs that might reduce the risk of heart attacks. As those opposed to using animals for product testing often remind us, such studies can sometimes yield misleading results, not least because the drug concerned has not necessarily been tested against a wide range of physiological and environmental conditions. Animals in the laboratory thus tend to become nothing else but their material bodies; the stressful conditions of their lives (overcrowded cages, say) are unwanted sources of variance in an experimental design. Not surprisingly, these data often bear little relationship to the situations in which many people will be at risk of heart attacks, or will subsequently take the drug.

A third tension between the socially constructed self and the mindless body-machines of science shows in the construction of animals and their bodies as lacking sexuality or desire; all they seem to have is 'sex' or 'reproduction'. In this sense, too, their bodies are pure materiality, lacking the power of expression that is part of human desire. Their bodies are sexed through possession of particular anatomies, but our bodies are at least allowed sexuality, even in the pages of scientific books. In part, this emphasis in biological thinking derives from the notion of reproductive success, according to which an animal is said to be biologically fit if it succeeds in passing on its genes to the next generation. But this idea, however central to scientific thinking, itself depends upon a cultural context that sees sexuality primarily (if not entirely) in terms of reproduction.

Within that limited framework, animal sex is equated with heterosexual mating (which can lead to reproduction). No other expression of what, in humans, might be termed sexuality, enters the definitions of animal sex; it becomes, by default, deviant. The sexuality of other species in scientific narratives collapses onto a breeding imperative. Certainly anatomical accounts refer to 'the reproductive system' rather than sexual structures; the word 'genital' is defined in the *Oxford English Dictionary* as 'pertaining to animal generation'. Thus in a book entitled *The Vertebrate Body*, we find descriptions of the hemipenes of snakes that can be 'inserted in the female cloaca to guide the sperm'. An inevitable consequence of viewing everything through the prism of reproduction follows the description of penile structure: 'These structures appear in the embryo in the sexually indifferent stage, and persist in the adult female in a relatively undeveloped and functionless state as the *clitoris*' (this is the sole reference to anatomy of females in a section entitled 'External Genitalia'; Romer 1970: 390).

What is important here is not the veracity of the statement but the way in which it is written and the assumptions it contains. The reference to the clitoris comes at the end of a paragraph on reptiles (I doubt that any biologists have even tried to study the possible effects of the female's clitoris on reptilian behaviour, however), but they establish the notion of the functionless clitoris. The subsequent paragraph on mammals makes only passing reference to the clitoris as being smaller than the penis (the subject of the rest of the paragraph). The mammalian – and thus human – clitoris remains functionless in this account.

Now none of this is surprising to feminist critics. My point here is to emphasize how 'sex' in non-human species is constructed; scientific accounts take for granted certain assumptions about sex/reproduction. The clitoris is functionless, with respect to reproduction, the reasoning goes. 'The animal body', then, typically lacks desire and sexuality, having only reproduction. In that sense, it conveys a notion of a body that differs from at least some accounts of human bodies; the human body desires, whether that desire is seen to be socially constructed or the product of raging hormones. But it is also similar, in that the sexed animal body has cultural meanings akin to those of the sexed human body. Both are assigned to one or other sex, male or female, largely on the basis of anatomy. If the anatomy cannot be relied upon to make the separation, then reductionist science must have recourse to some other kind of explanation. For example, the spotted hyena is unusual among mammals in that the external genitals of male and female look remarkably similar. Reproductive difference, however, is central to the way that biologists think about animals such as hyenas. So, they have sought explanation in the internal physiology and evolutionary history of these animals. Females, they speculate, are exposed to high levels of androgens (so-called male hormones) before their birth. These hormones then 'masculinize' the genitals permanently. But, even if scientists have trouble, the animals apparently can tell the difference and mating carries on as usual, and the biologists can rest assured that the tale of two sexes is saved. Sexual difference, whether human or not, is paramount in nature stories; it is the basic scientific 'fact' on which tales of animal exploits are built.

Writing about how we understand the concept of sexual difference in humans, Moira Gatens (1992) emphasizes that difference has more to do with how our culture marks bodies than it has to do with an underlying biology; experiencing our bodies may have widely different meanings for different women in different contexts. The widespread assumption that anatomy somehow provides the basis is, she insists, itself a cultural product:

> it is a particular culture which chooses to represent bodies anatomically . . . the human body is always a signified body and as such cannot be understood as a 'neutral object' upon which science may construct 'true' discourses.
>
> (Gatens 1992: 131–2)

Indeed, but neither is 'the' animal body. What we take to be significant differences between ourselves and other animals are themselves culturally coded, they are signifiers of a culture that chooses to represent bodies anatomically, and animals as inferior beings. It is differences *from* the human, I would argue, that bear the weight of cultural representations; bodily similarities between animals and us are much less potent.

Understanding 'the' body

The oppositions between sex and gender, between biological determinism and social constructionism present problems for feminism. At least part of the problem is the conceptualizing of *the* body. This serves to essentialize: '*The* body connotes the abstract, the categorical, the generic, the scientific, the unlocalizable, the metaphysical' (Fuss 1989: 52; emphasis in original). We may not be able to imagine 'the' body, but we can certainly imagine, as Adrienne Rich notes, my body (see Fuss 1989: 52). It is my body that is part of who I am.

If we still tend to think of 'the' body, then this is at least partly the heritage of the way that science has constructed the idea. With the development of detailed anatomical studies in the eighteenth century, 'the' body became a specific site for investigation; it thus acquired a specific meaning in relation to the development of biomedical discourse. And, of course, 'the' body became overlaid with culturally laden meanings of gender. So, for example, the structure of the skeleton became assimilated, during the eighteenth century, to the growing concerns in science with gender; where once 'the human skeleton' was depicted, it increasingly became a *sexed* skeleton (Schiebinger 1989). Animal bodies, too, acquire the definitive article and become overlain with meaning; thus we can find for example atlases of *the* mouse brain, or descriptions of the biology (for which, read anatomy and physiology – the bodily functions) of *the* rat. 'The' body in this context is inscribed with meanings of a species – the rat body is one that stands for the species rat.

Both liberal and Marxist thought, as heirs of the liberal humanist traditions, have failed to deal with bodies as anything other than biological 'facts', and it is within these that 'the animal body' remains. Moira Gatens (1992) points out how both the 'somatophobia' that feminists have noted in western thought and the celebration of the body in some feminist writing share an underlying understanding of 'the body as a given biological entity which has or does not have certain ahistorical characteristics and capacities' (129).

Yet to speak of women's biology is not to speak of anything outside history. As Ruth Hubbard has noted, 'Women's biology is a social construct and a political concept, not a scientific one' – that is, it is not an ahistorical, contextless given. Whatever constitutes women's biology at any point in history is a complex product of social/cultural construction (for example, in the way that medicine has produced notions of what women's biology is/should be), of 'biological' factors (such as the genes we inherit from our parents), and of the way those factors are affected by our immediate environment (it is not only because of my genes that I am relatively short; it is also, no doubt, a product of beginning my life amidst the rationing of post-war London).

What Hubbard is emphasizing here is that differences are complexly produced. There is, as we have seen, a tendency to locate whatever appears to be relatively unchanging (such as 'sex' differences in anatomy at birth) within something called biology, while whatever is more labile (such as 'gender') we are inclined to attribute to a (human) sociocultural history. Yet there is plenty of evidence (even from that much-contested terrain, the discipline of biology) that this dichotomy fails us. If, in our developmental histories, something appears to be relatively fixed, why should we assume that it is to do with biology? Why, too, assume that things that seem to be more variable are automatically the product of social factors? And why not assume that, if we perceive something to be unchanging, it is socioculturally produced, while something more variable may be related to 'biological' factors?

Indeed, it is from within feminist thinking itself that we can find an example of theorizing that attributes apparent unchangingness to socio-cultural factors. In the early days of the current wave of feminism, theorists emphasized gender relations as labile, as socially conditioned. But, other theorists began to point out, this failed to account for the ways in which gender was so consistently produced from generation to generation. If it was solely social conditioning, then it ought to be more susceptible to change, they reasoned. This was the basis of theories founded on the recurring patterns of childrearing (that is, that childrearing is done by mothers) that, various theorists noted, reproduced particular psychologies and hence reproduced gender divisions in particular forms (e.g. Chodorow 1979).

We are, however, more resistant to thinking about lability as having much to do with biology. If something is variable about humans, then we assume that it must be something we have learned or acquired. But many of the 'sex differences' in physiological functioning that are de-scribed in textbooks are differences that, in humans, are highly culturally contingent. Strenuous exercise, for example, is important in the develop-ment of particular characteristics (of heart and blood, say); thus, alleged sex differences are considerably less or nonexistent between trained athletes than they are in the rest of the population (see Birke 1992).

Bodies change. The belief in fixity belies this obvious statement, and encourages us to forget how our bodies are shaped *in* our culture. Few of us can completely ignore the powerful cultural messages that we should be thin to be healthy, for example. Yet to maintain thinness often involves struggling against the odds, to impose an apparent 'fixity' on ourselves. The processes of ageing too, are about change. We tend, in our culture, to reject these changes, to try to find ways of pretending that we are forever young, forever the same. But our bodies change, nonetheless.

If that is so for humans, then we should expect it to be the case for non-humans. But we should not forget that the constraints of scientific methods themselves impose a lack of variance in laboratory studies. So the knowledge gained from science about non-human species must inevitably lead to images of the ways that they function as stereotyped or fixed; experimental designs have tried to obliterate all those other sources of difference. Laboratory rats must be all the same size in an experiment, all roughly of the same weight, of the same age. Indeed, given the monotonous and impoverished conditions that laboratory animals must often endure, it is no small wonder that they behave at all.

To the extent, then, that 'the biological' has been theorized at all in feminism, it has largely focused on the meanings our culture gives to (human) bodies. But it has still tended to see 'the biological' as ahistorical and fixed, a kind of presocial bedrock. The physiological processes of the body seem to be left out of the social constructionist universe. This tendency to see the biological as somehow outside of culture applies also to how we see other kinds of animals; indeed, it informs how we construct our ideas of 'the animal' (contra humans) in the first place.

This opposition may be explicit, but is more commonly implicit, through a kind of hidden opposition to humans. For instance, Code (1983: 546; cited in Hekman 1990: 143) argued that 'Human beings are creatures of a sort whose nature is, in large measure, *structured* by nurture'. Indeed, but a subtext of this claim is that there are other creatures, whose nature is not structured by nurture – other kinds of animal. Such an animal – a product of a fixed nature – is critical to biological determinism. Explicit or implicit, its existence ensures that determinist notions continue to flourish. But such an animal and its body is, like the monsters of medieval cosmology, purely a fiction.

10

THE RENAMING OF
THE SHREW

Shrew: Any of the small insectivorous mammals belonging to the
genus *Sorex* . . . [but also] malignant being . . . wicked, evil-disposed
. . . a woman given to scolding or other perverse or malignant behaviour . . . a scolding or turbulent wife. . . . [as verb] to curse . . . to
deprave.

(*Oxford English Dictionary* 1978, Vol. IX: 773)

They noted, that although the virgin were somwhat shrewishe at
the first, yet in time she myght become a sheepe.

(Lyly, 1580; *OED* 1978, Vol. IX: 'Sheep'; 5, b)

This is a way to kill a wife with kindness,
And thus I'll curb her mad and headstrong humour.
He that knows better how to tame a shrew,
Now let him speak; 'tis charity to show.

(Shakespeare, *The Taming of the Shrew*, Act IV, Scene I)

In this book, I have tried to explore several interconnecting themes that
centre on how we see animals, and the relationship between humans
and other animals; I have approached this task as a feminist and a
scientist, concerned to think about the kinds of questions that feminists
might ask about science and animals. The shrew stands as an icon in this
task both for its ambivalent meanings to women, and as representing
the way that science has (re)named animals of all kinds.

My interest as a feminist (with a 'mad and headstrong humour') in
exploring and critiquing the biology that I have studied over the years
joins forces here with my passionate caring about non-humans and how
we treat them. Ultimately, what concerns me is the profound lack of

respectfulness that our culture has developed towards others – be they other kinds of humans, other kinds of animals, or other parts of nature.

Feminism has documented countless examples of ways in which different groups of people are cast as other and treated with less (or no) respect; the animal epithets thrown at uppity women are one manifestation of that. Shrewish women had to be tamed, to become sheep; neither description implies respect (either to women or to the animal kinds invoked). 'Othering' is typically founded upon perceived difference from the dominant group of able-bodied white heterosexual men. So, too, are non-humans and nature cast as 'others' – always different and always lesser. Although, sadly, I think it is almost impossible to achieve change in science without substantive (or revolutionary) change in our society, we do need to break down the boundaries by which we have come to see those 'others'. We need to develop a greater respect for all creatures – indeed, for inanimate nature, too.

In trying to seek interconnections I have raised many questions. Here, by way of bringing some of these threads together, I want to focus on some broad themes that seem to me to need further development, both within feminist theorizing and outside of it. These include: the relationship between feminist thinking and animal rights; empathy and caring; what questions we might ask about animals in relation to feminist debates about science and epistemology; the further development of ideas about non-human animals as autonomous creatures with needs broadly similar to our own; and, finally, how we might tell different stories about women, animals and nature.

A central motif in this book is the concept of 'the animal', and the question of why we in western culture seek to distance ourselves from it – not least by defining it as a beast lacking any finer feelings or civilized reason. Modern feminism has grown up with the heritage of the liberal traditions of the Enlightenment, and the views of nature that have developed alongside it; feminists, too, often seem to distance ourselves from this mythical creature (even while, ironically, many of us live with other animals). At a time when many of us are concerned about threats to the environment, to nature, that heritage is a millstone around the neck. For how can we ever learn to respect other creatures if we persist in naming them as somehow lacking? We may no longer name the shrew as evil and malevolent in western culture, but now we name it stupid, deficient, lacking in the social graces.

Beyond 'the animal'?

The concept of 'the animal' is undoubtedly a powerful one in our culture, and helps to define us. Separating ourselves from 'animals' becomes a statement of our sovereignty over nature. Although feminists have not often used this contrast, it certainly underlies the ways in which we think about biological determinism, as I have argued. Animals' bodies and lives remain mired in the bogs of pure biology, while human minds miraculously make the escape.

One reason why we have relied on these dichotomies is that the study of animals (with their object status) has been defined as part of biology. Humans are contradictorily both objects (studied as such in the human sciences) and subjects, creators of knowledge, in ways that, by definition, animals cannot be. Yet the opposition between human and animal goes deeper than simply the definitions of what is, or is not, included in any disciplines of natural science. Why is it that we defend and police these boundaries so astutely?

An opposition between nature and culture emerged most strongly in the West during the Enlightenment, the period in which the study of 'man' developed, and the 'human sciences' were born (Horigan 1990; Benton 1991). Once separated, the boundaries now had to be policed to ensure that the 'study of man' remained unsullied. Those creatures that threatened the boundary – the 'savage', feral children, animals – posed particular problems for demarcating nature from culture. The solution was twofold: some of these creatures were categorized as falling exclusively into the class 'nature', such as animals, while the very existence of others (such as feral children) was increasingly contested (Horigan 1990: 7).

While the emerging human sciences were defending their territory and creating an intellectual space for themselves, the natural sciences came more and more not just to study 'nature', but to adopt a particular philosophical position – reductionism – in the service of that study. The success of molecular biology in the late twentieth century has served only to consolidate that position, and thus to drive the wedge between human and natural sciences even deeper (Benton 1991). The boundaries are policed (and suspiciously) from both sides.

'The animal', as it emerges from *both* studies of human society/culture, and of non-humans in the context of biological sciences, is deprived of just about everything except base instincts. Its capabilities become reduced to a form of evolutionarily-selected adaptation to its environment that allows little room for inventive thought, for intuitions, for creativity. It certainly is not like the animals with whom I share my life at home.

For feminists critiquing science, the very existence of this mythical beast is a problem, for it helps to reconstruct the nature/culture boundary. It *is* nature, pure biology. In turn, those who would defend a belief that there is a little bit of this creature in all of us use tales of the mythical beast as a victim of its genes to defend their beliefs. And so the fables of biological determinism – of animals as models for fixity in our own behaviour – persist.

That, at least, is one important reason why feminists should be concerned about 'the animal'. But not only as the concept, important though that is to feminist critiques of biology; there are important reasons why we should also be concerned about *animals themselves*, about the fate of other (non-human) beings. We have engaged, in our writing and political practice, with struggles against injustices of all kinds, whether that is on grounds of race, class, sexual orientation, having or not having an able body, as well as gender. In that sense, feminists have challenged several systems of domination, believing as we did so that these oppressions are deeply intertwined.

Feminist critics of science, meanwhile, have stressed how the development of western science and technology has gone hand in hand with the domination and expropriation of nature. If they are part of that nature, then it seems to me that the domination within our culture of non-human animals matters; it matters because systems of domination remain intertwined, so to understand human oppressions more fully requires us to consider also how those oppressions are related to nature's domination – and to struggle against them all. Nothing less than the future of global environments is at stake.

Yet other animals are not *in* nature in a way that we are not; many of them are clearly very much *in* human culture and society. 'Nature' is not something 'out there', divorced from human experiencing. To follow through the argument that humans are very much 'in' nature, and animals can be thought of as in culture, then humans and some non-humans begin to occupy much the same terrain. To argue that humans must always come first, that human oppressions must be fought before that of (say) cats or dogs on grounds of loyalty to our own species is to relegate the animals to a subordinated category of nature (as well as reifying the concept of species). In so doing, it also denies our own place in nature. But if we dissolve or dislocate the nature/culture boundary, what then?

I am not convinced by the argument that there is only so much time and money in the world and therefore we should concentrate our efforts on alleviating those injustices against fellow humans. Moreover, not only am I not convinced by that but it seems to me that a great many of the injustices that humans perpetrate against animals are themselves deeply embedded in the very same systems of domination that lead to injustices against humans. It is very easy to see injustices against animals as simply the manifestation of individual levels of cruelty but I don't think that is always the case. Some undoubtedly are, but many are embedded in those forms of domination. The ways in which animals are treated in the laboratory is, in part, a product of the location of science in the world of late twentieth century industrialized capitalism. Our culture, after all, requires animals for testing new products; whether or not those new products are necessary in any real sense of human need, what happens is that those animals are used to test the products for safety in relationship to the profit margins of the companies that are producing them.

Feminism, science and animal rights

There is a growing antipathy towards the use of animals in scientific experiments, alongside burgeoning interest in environmental protection. The arguments used by feminists to challenge the assumptions scientists make about using animals are based primarily on the need to extend our critiques to include other systems of domination, and on appeals to empathy and understanding. Animals are fellow sufferers at the hands of science, and women, the argument goes, are more likely to empathize with the animals' plight.

Feminist involvement with the wider environment, with ecosystems as a whole, raises similar issues about the relationship of beings of all kinds, plants, animals, humans, in relationship to each other. Many people are growing increasingly concerned about the ways in which those ecosystems interrelate, the ways in which those fragile relationships are being damaged by human action. But it is also about human action in the pursuit of greed and profit; the destruction of the ozone layer and growing problems with global warming are a product of the destructive capacity of industrialized society. A concern about the relationship of humans to non-humans has to be a part of that.

However tricky it is to link women and nature, feminism and animals, the argument that women and animals are often fellow sufferers is a powerful one. It stands, moreover, in stark contrast to the appeals to rationality that underpin the writings of several male philosophers about animal rights. Whatever the appeal of ideas of 'animal rights' (and there are many), the whole concept of rights is problematic – not only with regard to animals. This is not to reject out of hand the spirit behind thinking about rights; rather, if there are problems with the concept of rights (especially from a feminist perspective) then there will also be problems applying them to non-humans.

The concept of rights has arisen primarily within the liberal–individualist discourse of the West. It seeks to preserve personal autonomy, and assumes that the individual herself is the best judge of personal interests. However, as Benton (1993) points out, this ideal does not easily transfer to the case of non-human animals, whose interests in human culture (which is surely part of the point at issue) would have to be judged and argued for *by* humans. We may see ourselves as advocates for animals, but they must be the best judges of their own interests.

Perhaps more importantly, the tradition of liberal rights does, in principle, grant protections to all kinds of people, and even in some contexts to animals. But these are protections that 'they cannot in fact deliver under prevailing conditions of highly concentrated property in productive resources, social and economic inequality, and market-generated relations of mutual antagonism and estrangement' (Benton 1993: 198). Women, after all, have some protections under the law. In the context of patriarchal society, however, I would hardly claim that women in Britain can universally exercise even those 'rights' granted us under the law. After 15 years of right-wing Conservative government in Britain, women have had more rights eroded than granted.

The law could hardly be said to offer rights to laboratory animals (in the positive sense of 'rights' implied by the animal rights movement). It does attempt to offer some degree of protection against abuses, which is at least part of the rights picture, but it will always, ultimately, fail to achieve even that, just because science is deeply embedded in the social and economic conditions to which Benton refers. Those contexts will ultimately be far more powerful in determining outcomes than the interests of the animals, however well protected they may be (against some abuses) within law. Some individual scientists may perhaps have a cruel streak; most do not in my experience.

What drives many of the abuses of animals in science is that wider context – the profit margins of commercial companies testing cosmetics; competition between companies; the pressures to get a large grant to fund research; the social pressure to conform to an ideal of unemotional detachment. Sarah Kite's descriptions (1990) in her undercover work in a laboratory for the British Union for the Abolition of Vivisection are harrowing; her tales of suffering endured by beagle dogs must surely touch readers' emotions. But what she witnessed was animals used for toxicological testing – testing which is driven by commercial gain and by public demand for yet more (safety-tested) products. Even the minimal protections granted in principle by the law can have little weight against such commercial demands.

Feminism itself has depended in part upon concepts of rights, whatever the problems these present. We spoke, for example, of 'a woman's right to define her own sexuality' in the women's liberation movement of the 1970s, or her rights to financial independence from men. But that was to take the individual out of her context; 'rights' to define one's own sexuality became increasingly questioned in the 1980s, with debates about sadomasochism and violence against women in pornography. The difficulties with 'rights' and their origins in liberal individualism apply just as well to the case of animals; feminist writers have also challenged the kinds of claims made by male advocates of animal rights.

The primary problem for feminist critics has been the reliance in the animal rights literature on rationality and a justice conception of rights. This is clearly enunciated in, for instance, Singer (1975), Regan (1983) and Rollin (1992). Rights that depend upon rationality are tricky for feminists, for how often have women been denied rights on the grounds of our alleged irrationality? Now the denial of rationality has shifted (somewhat) from women, onto non-human animals, but here too it is tricky.

The animal rights position seeks to rescue at least some kinds of animals from that abyss of being declared forever irrational and stupid. I could perhaps argue for the rationality of other mammals, based on what I know from various sources about their capabilities. I would be pushed to argue rationality for, say, some invertebrates – not because they lack rationality, but because I do not know how we could ever know if they had it. This is a problem for, say, Tom Regan's concept of rights based on some animals having interests, as subjects of a life; it seems to require that we draw lines somewhere between those having rationality/consciousness (who therefore can have interests) and those without. I just do not know how *we* can decide who in the non-human world can have interests, self-consciousness, or powers of reason; it is too close for comfort to histories of men deciding whether or not women (or non-Europeans, and so on) can be declared as possessing such attributes. Besides, how would we know whether animals do – particularly if we define the capacity to have interests or self-consciousness in a way that excludes most of them?

The rights/justice position relies, too, on individualism, and expands the circle to include non-humans on the basis of sameness, or at least similarity; it thus ignores or downplays difference (Slicer 1991). In doing so, it is fundamentally anthropocentric – what is valued in this view is similarity *to* humans. If difference is valued at all, it is difference *from* humans. There is no celebration here of the uniqueness of each and every kind of animal, nor even of their differences one from another. What matters is what they share with us, allowing us to name some of them as honorary humans for the purpose of moral accounting. This is what encourages the use of animals with nervous systems not like ours ('simpler' ones) in science, in preference to those with nervous systems similar to ours ('higher' animals).

One strand of animal rights arguments draws on utilitarianism, which seeks to weigh interests in order to determine the greater good. In arguing for animal rights, Peter Singer uses a utilitarian calculus based on the ability to feel; many animals can feel, and so can suffer, he argues, which give them interests that should be taken into account rather than ignored (Singer 1975). Animal rights, in this context, then becomes the rights of sentient creatures to have their interests taken seriously in any balancing against the interests of other creatures.

Utilitarianism does allow the possibility that animals might feel or suffer. But it is still locked into rationalist decision-making around interests, as Josephine Donovan suggests (Donovan 1990). This is particularly, and ironically, clear in relation to the regulation of animal experimentation in Britain. Utilitarianism is explicitly enshrined in British law governing the use of animals in experiments, as we have seen. It requires that potential benefits (the good of humans, in particular) be weighed against potential suffering of the animals. In that case, it is a utilitarianism employed in the service of science, relying explicitly on a discourse and process rooted in rationalism.

What the utilitarian calculus seems to gloss over are the important questions of who decides costs/benefits? which animals? which humans? Animals may stand to benefit in the case of some veterinary research,[1] but more often, the lives of laboratory mice are weighed against vague claims about, say, possible gains in understanding of 'how cancer develops' or 'the aetiology of AIDS'. It is obviously humans that must decide on the balance – and who most clearly will benefit. Humans (and not all humans at that) have made the rules, interpret them, and create the narratives. The mice don't write the stories.

If they did, I have little doubt that mice would make different decisions about the greater good. But flippancy aside, the issue of who makes the utilitarian decisions is an important one. If it is not the animals who decide, neither is it most humans. In general, the decision-making of this kind of practical utilitarianism is done by men, and by men from privileged class and race backgrounds; deciding the 'greater good' in the design of a scientific experiment is not a democratic process, open to all. So whose 'good' is it likely to be?

Scientists are encouraged in this process to consider whether a 'simpler' species might be used to achieve similar ends, so shifting the balance point (simpler species, we must infer, are going to suffer less). The question of which animals hinges then on the kinds of stories that scientists and philosophers tell about the capabilities of different kinds of animals. Scientists have much to gain by defining some species as 'simple'; in Britain, for example, the law requires them to use simpler species if possible, so a research project could stand or fall on the basis of species choices. Defining animals is, moreover, part of the work of scientists, whether that be classifying them or categorizing them according to the structure of their nervous systems; yet it is scientists whose research directions may depend upon the outcome of that classification. The stories that science tells are never neutral.

One of the strongest criticisms from feminists of the animal rights/

rationality position is that it excludes emotionality or caring, thereby excluding an important arena for feminist analyses (Donovan 1990; Slicer 1991). Josephine Donovan does not feel that the arguments in favour of animal rights go far enough for feminists. Both Regan and Singer, she notes, stress the need to avoid sentimentality and emotionality in advocating rights for animals. Both privilege rationality – yet rationality both denies the significance of empathy (too emotional) and supports the very methodologies of science that exploit animals in the first place. This point echoes a wider feminist debate centring on the problems for women/feminism of relying solely on principles of justice to make moral claims; for many feminists, particularly following the work of Carol Gilligan (1982), what this leaves out is the principle of caring based on affectional bonds.

Caring and empathy

As I noted in Part II, empathy for animals and sensitivity to their plight is not encouraged in the training of laboratory scientists. It is also thought 'unmanly', echoing the masculinity of laboratory culture with its claims of distancing objectivity.

There are two related dangers in thinking about empathy and animals, however. The first is one that has beset several feminists writing about the gender of science. To say that science is gendered as masculine is to make claims about how science is practised, about how scientists see the world (Keller 1985). Unfortunately, it is all too often interpreted as being a claim about who does science. So, in relation to thinking about animals, this would collapse into a claim that women uniquely have special abilities to empathize with animals. Some may, but that is not a claim I would necessarily wish to defend. The second problem is that the trait that becomes associated with women – empathy – is taken to be intrinsically associated with women's reproductive role. Women thus become nurturant or empathic because they are biologically designed to be mothers.

In suggesting that empathy and respect for nature may have overtones of gender, I am saying that these are qualities that have become stereotyped in western culture as feminine. That is not to say, however, that they are the prerogative of women. I think these are qualities that contribute to being a good scientist, whatever the sex of the practitioner, contributing to a 'feeling for the organism' as Evelyn Fox Keller put it in her biography of Barbara McClintock (Keller 1983).

Some feminist writers, moreover, have argued that women may indeed develop an 'ethic of caring' out of their experiences of doing the emotional work of servicing others. Thus, Sara Ruddick suggested, for instance, that women might develop a special kind of thinking, maternal thinking, in the course of their work in mothering (Ruddick 1982), while Hilary Rose (1983; 1994) has noted how such emotional labour forms part of women's experience in relation to science just as it does elsewhere. If caring forms part of women's collective experience, then caring about nature (including empathy for animals) does begin to seem the prerogative of women.

Other feminists have voiced doubts about founding a feminist politics on an ethic of caring, however (Houston 1987; Hekman 1990). There are two important reasons, critics urge, why we should be wary of *basing* a politics on such 'womanly' virtues. The first is that it is difficult to see how we could advocate a women's standpoint and politics when women are themselves so divided (Harding 1986); certainly, not all women's experiences are shaped through nurturing children (including my own, as a child-free lesbian). The second problem for critics is that elevating women's virtues so often seems to rely on essentialism, as though women were inherently (biologically) nurturant, peaceful, or whatever. One of the most bellicose prime ministers in recent British history was, after all, a woman.

I am not convinced that essentialism is a necessary consequence of arguing that at least some women might share experiences of caring and empathy from their lives as women. Some of these qualities might, in some situations, simply be expressed more easily by women because it is socially permitted. I noted earlier, for instance, how some scientists felt that female animal technicians might be better at handling animals, perhaps because of greater empathy or perhaps simply because it is culturally more acceptable for women to show these qualities. Caring and empathy may, at times, be more likely to be found among women, but what I think is more relevant to my arguments here is that these are qualities that bear the hallmark of gendered stereotyping in our culture and that are also stereotypically not encouraged among trainee scientists.

In that sense, there is dissonance, which individual women and men may experience to different degrees (while some don't see empathy in science as having anything at all to do with gender constructs). Empathy and respect for animals as sentient, thinking creatures with lives of their own are not substantially part of the written narratives of science. In laboratory culture, too, there is often a kind of macho heroism that seems to fly in the face of any respectful stance. If we are to change cultural perceptions of both women and nature, if we are to change the position of women who want to study science, then there will have to be careful attention paid to the denial of empathy in the process of becoming a scientist.

Developing empathy and respect in science must go hand in hand with developing a caring and relational approach to nature. Val Plumwood (1991) notes the strength in western culture of the self/other motif, and its relationship to rationality and instrumentalism. This self defines itself against others, and denies connections with them. Plumwood urges instead that we use feminist insights derived from the work of Carol Gilligan and others, that suggest that women might tend to have distinctive 'ways of knowing' (see also Belenky *et al.* 1986). These ways of knowing are relational, and emphasize embeddedness. Again, Plumwood is not saying that these stances are the prerogative of women. What is at issue is that understandings based on relational qualities might be gendered; women, because of a gendered division of labour, are on the whole more likely to have these qualities.

If empathy for animals in science is indeed gendered, as I have argued, then we have two important tasks. The first, most obvious one, is that we have to deconstruct that genderization and its implicit values. I want to see a science that is based on respect for all the creatures it studies, on acknowledgement of their sentiency and potential for suffering; laboratory machismo, not showing the emotions, must be challenged and shown to be the myth it is. Indeed, it is more than a myth; it is, I would argue, detrimental to good science. Understanding how nature works does not seem to me to follow easily from disrespectful dismemberment.

The second task is to alter the narratives. Many scientists do, in conversation, speak in terms of animals as active knowing subjects. Yet we are all trained to deny that knowledge in writing, or as part of our public personae at conferences. Students of biological sciences are, in increasing numbers, posing challenges; why, many ask, should we be expected to do things to living animals, or dissect ones that have been specially killed? It seems to me to be profoundly dishonest of those of us who are scientifically trained to ignore the contradictions in our *own* thinking as scientists if we scoff at what these students say. More open recognition of ambivalence would go some way towards changing those narratives.

Feminist critiques of science

Feminist work around science has developed in several directions (see Rosser 1989), some of which at least touch on my themes here. The clearest example is the feminist critiques of biological determinism, which I addressed in previous chapters. Another strand of feminist work around science has focused on equity issues – why are there so few women in science, now or in the past? Here too there are questions to ask about how we see animals/nature.

One of the many ways that question can be addressed is to look at how women's perceptions of themselves may be at odds with scientific narratives. While both women and nature are seen as the objects of scientific enquiry, there is a problem for a woman entering the sciences, for how can she be both object and objective scientist? Gendered expectations around empathy versus distancing, too, may well contribute to the way many women feel science is not for them. How many, I wonder, have dropped out of biology because they did not want to dissect an animal?

Another major area of feminist criticism has focused on biology, criticizing the determinism and reductionism of much biological research. Increasingly, the study of biology has become synonymous with the study of molecules – the explosion of research into genetics and DNA is a clear example. Within this kind of reductionist logic, the organism becomes seen as little more than its component parts. There is not much room for respect here.

Not only does reductionism then encourage us to see those parts as causal (seeing hormones as causing behaviour, for instance), but it also allows dismemberment. For if the processes that biologists study are collapsed mechanistically onto the component molecules, then to study the processes we must get at the molecules that drive the machine. To do so, in turn, requires that we take the machine apart. Granted, scientific practice in the late twentieth century is not so crudely based on imagery of the beast-machine operating like a clockwork mechanism as it was in earlier times; today, at least, scientists would expect to anaesthetize the animal first. But the metaphor lingers on, and taking the animal apart is undoubtedly assumed to be a crucial part of the process of 'getting at' the component parts.

Feminist critics have emphasized the political context of science. Science, we have often noted, has developed in a society which makes virtue out of greed and profit; not surprisingly, in that context, the practice of science itself is driven more by profit margins and the political agendas of funding agencies than it is by people's needs. Not surprisingly, the needs of non-humans (and of many humans) are not important to scientific agendas. While it may be problematic to think about the possibility of a 'feminist science' as long as we live in a non-feminist society (Longino 1989), we can nonetheless speculate on what would need to change; one aspect of science that, I believe, must be challenged by ideas of feminist science is the ways in which science uses animals (see Birke 1986).

An obvious example of scientific use of animals in this vein is in testing 'inessential' products, such as most cosmetics. These products do not usually satisfy a basic need (though they may certainly satisfy 'needs' that come from the creation of desires in a consumer society). Indeed, most basic human needs – for shelter, warmth, food and so on – are needs that we have in common with other animals (Doyal and Gough 1991). So if we are seriously to consider how feminists might challenge science, then the abuse of sentient animals in the pursuit of goals that are inessential to human need or welfare must be challenged. For how can we justify denying basic needs to other creatures in order to satisfy a desire in humans? Any justification seems harder still when we remember that such desire is very far from the satisfaction of basic needs for many humans throughout the world.

Cosmetic testing is a relatively easy case to challenge, as many commercial companies now claim to market products that have 'not been tested on animals'. Whatever the merits of those claims (and some would contest them), it is clear that many consumers seek those products. But what about animal experimentation in cases where it is claimed that the research or product is *essential* to human needs? Or even to animal needs, as in the case of veterinary products?

Here the story is much less clear. If we can conceive of a science that is genuinely 'for the people' then might we imagine animal-based research that was driven by such altruism? Cancer research is often quoted in this context, and defenders of animal research frequently emphasize the eradication of major infectious diseases such as smallpox through vaccination programmes (based, we must infer, on animal research). I have no doubt that some use of animals by scientists is indeed motivated by a desire to help eradicate human or animal suffering; I think it likely, too, that vaccination did help to control diseases, and that many important developments in clinical medicine have been helped by experiments using animals.

Yet I have also seen animals treated in laboratories in ways that were far from respectful (or were downright cruel); I know, too, that some of the claims made in defence of animal use are overblown. The decline in, for example, cholera and tuberculosis owed much to changes in living standards. For those reasons, at least, I am somewhat sceptical of the overly enthusiastic claims sometimes made on behalf of animal research. But there are other reasons, given my arguments in this book, for questioning the use of animals *even* in the case of research that claimed to rest on meeting human needs. Most importantly, to challenge the twin assumptions that animals are in nature, and that domination of nature is intrinsic to human culture, means to call into question the ethical grounds for using animals as 'models' for humans. If at least some animals are in culture, then how can we justify using them in ways that we would not usually use fellow humans? What does this say about the nature of the scientific knowledge that is created, if it is premised on such contradictory relationships to non-humans?

Animals, feminism and scientific knowledge

Animals have, of course, been central to the construction of scientific knowledge; they have been named, described, dismembered and disfigured in the name of that knowledge creation. Here, however, I want to ask how feminist approaches to scientific knowledge might engage with questions about our relationships to animals.

A major strand of feminist work around science has indeed focused on the nature of scientific knowledge (although not explicitly about animals). Sandra Harding, in *The Science Question in Feminism* (1986), distinguished three perspectives in feminist writing about science, each posing different epistemological questions. The first of these is what she terms feminist empiricism, a stance which starts from the premise that much of what feminists have criticized is 'bad science', and arises from androcentric biases built into the research. This is indeed something feminist critics have had to do at times, and as political strategy it has its uses (it has recently been part of the way feminist biologists, among others, have publicly criticized the claims made about 'homosexual brains'). Its draw-back, however, is that to speak of biases is to assume that we can have a bias-free, or value-free science – a claim which flies in the face of most feminist thinking about science.

The second broad strand that Harding identifies is feminist develop-ment of standpoint theories (see Hartsock 1983; Rose 1983; Collins 1991). These focus on how knowledge is created from experience, and par-ticularly from the experiences of being part of an oppressed group; in a society divided along gender lines, women will, on the whole, tend to have different life experiences from men, structured through their experience of gendered subordination. The value of standpoint theories lies in their feminist insistence on a shared epistemology born out of experience. They have, however, been criticized for their tendency to universalize, to assume too easily a common ground for 'women', so ignoring differences between us.

A third strand of feminist criticisms of science comes from post-modernism. Here, an important concern is to question any universalizing claim (or 'master narrative' such as positivist science), and to deconstruct boundaries (such as subject/object; or self/other; or the boundaries implied in concepts of social class). Donna Haraway's work, with its playful de-constructions, is one example in feminist analysis of science (e.g. Haraway 1990; 1991). Questioning claims to universal knowledge is important for feminism, and we have consistently questioned the societal belief in science as 'one true story'.

Yet the impulse toward postmodernism has been problematic, too, not least because feminism as a politics is necessarily founded on at least one kind of universality – the oppression of women. To emphasize and ele-vate difference seems at times to undercut the grounds for that politics. The postmodern insistence on partial visions, moreover, seems sometimes to lead us to a position in which no one account is more true/valid than any other; if so, then a feminist story is on an equal footing with a non-feminist story (which makes feminist postmodernism itself a contradiction in terms; if it is self-consciously feminist, then it must make positive assumptions about feminist stories; see Harding 1992).

Each of these three strands of feminist engagement with science offers something of value to thinking about animals. As will no doubt be evident, I have been influenced by, and draw upon, all three. Feminist empiricist approaches, for instance, remind us of the values built into scientific theories from the start – not only about gender (think of the homosexual rat, or the feminized brain), but also about animals and what they are. As I have argued, there are some ways of thinking about animals that seem to be 'better' than others, in the sense that they are more inclusive. So, even within an empiricist framework, I am more convinced by accounts that look at the complexities of (say) the development of young rats than by those that reduce their development to the ups and downs of hormones.

Standpoint theories can also shed light on how we think about women/nature. While I am concerned about how easily this collapses onto essentialism, I think there is an association between constructs of gender and how we think about nature; culturally, we think of nature in gendered ways. If so, then women collectively might well be more inclined to think of themselves as embedded in nature, rather than apart from it. There is certainly a strong strand of feminist writing which celebrates women's closeness to nature and her cycles. Feminist writing on an ethic of caring seems to me to be a useful background for thinking about empathy and its gendered denial in scientific training, but it is not without problems. An obvious one is that, however often I emphasize that associating empathy and gender does not mean collapsing it onto sex, some readers will read just that and see empathy as the preserve of (biological) women. A second problem follows on from that; seeing women as having special links to animals may break the woman/animal boundary down, but it also consolidates the categories of woman and animal, and consolidates them in nature, set against the universal 'Man'. Differences among us 'others' thus remain unarticulated and unexplored.

From postmodernism comes a concern to deconstruct boundaries through difference. Given the impact of these ideas on feminist scholarship, my interest in questioning the human/animal boundary is inevitably influenced by postmodernist thinking. My concern, too, with animals is not only with how we treat them, but also with how we construct our *ideas* about them – about, in short, the stories we write about animals. But postmodernist thinking seems to me to fail adequately to address issues of pain and suffering, of human cruelties towards animals (or those of men towards women). If the world is collapsed into a set of narratives, what happens to the lived experiences of non-humans? Power and domination, whether of humans over animals or between humans, are neglected in postmodern accounts.

In the political moves towards postmodernism, the self or subject of liberal humanism has come under attack; subjectivity itself is fractured. Subjectivity, of course, with its roots in self-conscious awareness, has been something we have long denied to animals. Yet that is now under fire; with the rise of animal rights advocacy, some animals are permitted to attain the status of subject (as in Regan's 'subjects-of-a-life').

It is ironic, however, that animals do so at a time when postmodernism celebrates the 'death of the subject'. Instead of seeing subjects as essentially given, this celebration draws on Foucault's insistence that 'subjects' are created through discourses; thus we move to a position in which the rational subject no longer exists as an autonomous entity. For feminists this has always been problematic – partly because power and oppression seem to dissolve into rhetoric and discourse (and that key phrase of women's liberation, the personal is political, seems to be denied) and partly because women have, traditionally, also been denied subjectivity. It is, moreover, the death of a subjectivity which has historically privileged consciousness (Gatens 1986). And consciousness is precisely that which has typically been denied to other animals in the same liberal traditions.

'The animal' as a beast without consciousness or rationality haunts western intellectual traditions. It can be found in the reductionist accounts of science; in the work of philosophers seeking to challenge claims of animal rights (Harrison 1989; Leahy 1991) and makes guest appearances in the texts of sociology. It is not, I have argued, the ghost of any animal that ever lived, but represents the spirit of our own existential anxieties about human 'specialness'. Those humans in positions of power (at least in western cultural traditions) tend to want to arrogate to themselves the qualities that they have constructed as being important. After the Enlightenment, those qualities happen to be things like rationality and speech.

Postmodernist approaches at least suggest the possibilities of deconstructing the boundaries erected between self-conscious 'Man' and other animals. Yet for all its attractiveness to feminist theorists, I have reservations about wholeheartedly taking postmodernism on board. The first comes from my own scepticism about the willingness of intellectuals to move far from their privileging of rationality and conscious thought. One consequence of this is that discourse is privileged in postmodernism as a practice founded in intellectual processes. Just because of the way humans (rather, male western intellectuals) have chosen to define it, this prioritizes what we claim only humans can do. In that sense, the impulses of postmodernism serve to reinforce the boundary between us and other animals.

Celebrating the death of 'the' animal, moreover, may not necessarily lead to any celebration of diversity and difference between different animals. I suspect they would remain in the quagmire of not-human, while postmodernism continues to focus on fracturing identities between humans.[2]

My second reservation is that postmodern fracturings seem so easily to deny power (at best, they relocate it into discourses in which power is constructed through all the participants). In thinking about nature/animals this does not work, at least for me. Science has contributed to the appropriation and exploitation of nature; in that sense, it embodies a system of domination, of human mastery *over* nature – and thus of power over animals. It is not they, but we, who determine the quality of most of their lives, and the time of their deaths. They participate in the discursive construction of that power only insofar as humans name them. Our power over them, moreover, extends to causing them hurt, real hurt, that does not disappear into webs of fancy narratives.[3]

A third reservation follows from that one: boundary transgressions may be exciting, but they may also have implications for the suffering of animals. The boundaries are beginning to be challenged in feminist writing. Donna Haraway's exploration of cyborgs is one example of how boundaries may be transgressed (Haraway, 1990); and from science it-self come all kinds of possibilities, in which organs, tissues or genes can be moved between organisms. As Emily Martin (1995) has noted, however, the very scientists who are developing the techniques by which the body can be broken down, or reassembled, may not themselves have taken in the implications of such boundary transgressions. Furthermore, the implications *for animals* may be profound. We are already witnessing suggestions that pigs be bred to supply hearts for human patients; and transgenic animals sometimes become factories for human use. These kinds of boundary transgressions are not ones I would wish to celebrate; it is the animals themselves that are the losers.

Renaming the shrew

How we see other animals is undoubtedly a human construction; we cannot know how they might see us (as hopelessly fragmented, per-haps?) In 'She Unnames Them', from which I quoted at the beginning of Chapter 2, Ursula LeGuin remarks upon the 'perfect indifference' with which the non-humans regard the naming that we humans confer upon them. Elderly female yaks did discuss the question of naming/unnaming in the story, but 'finally agreed that though the name might be useful to others, it was so redundant from the yak point of view that they never spoke it themselves, and hence might as well dispense with it' (1987a: 194).

In our naming, however, and the ways in which we construct our ideas about each kind of animal, we build the names and the animals into our lives. Creating those discourses has been a critical part of human cultures, and each culture has done it differently; yet there are common themes. Many cultures have included some kinds of animals *in* culture as domesticated; many have categorized some as potential food, others as non-food or as sacred; many have mythologies that tell of fabulous hybrid creatures, or of a time when humans and animals mingled and spoke one to another.

Because we have, in the late twentieth century, become familiar with the distancing discourses of science, we tend to forget that 'what we know' about each kind of animal is still a story. Consider what we know about shrews, for example. We might read in scientific prose, for example, that

The Common Shrew can be recognized by its three-coloured coat . . .
It [can be distinguished from related shrews] by very careful measurement of skull proportions or by examination of chromosomes.
(Corbet and Ovenden 1980)

These are 'the facts': the scientific account of the common shrew. We are expected to see this account as somehow more true than previous accounts; it was lack of scientific understanding, on this view, that led to the classification of shrews in the category of mouse, *Mus*, as the first edition of the *Encyclopedia Brittanica* did in the eighteenth century. It was an 'old wives' tale', not science, we must infer, that led our predecessors to see shrews as dangerous, liable to make horses lame. It was fanciful tales that saw the stereotype of the nagging wife in the squeaks of the tiny mammal.

Yet they are all stories, accounts of chromosome numbers included. What matters is that today we expect to believe the stories we label as scientific, and not to believe those we label in our culture as folklore. In emphasizing this, I am not trying to claim higher status for some of the folklore, or to claim that, say, chromosomes do not exist; I am saying only that they are constructed as different kinds of stories. I do not see shrews as dangerous, nor as capable of laming horses, any more than I see snakes as deceitful apple-fanciers. What I do want, however, is that we look critically at the scientific stories, at what they in turn convey. One thing that strikes me about the folklore is that it often portrays a rather more active, thoughtful kind of animal than the abstract accounts of science and chromosome numbers ever do; it may have been cunning and dangerous, but the medieval shrew at least had a life of its own.

Not so the animal in the scientific accounts, who often seems to disappear into the passive language and the (human/male) gaze of the scientist. The scientific account may be 'more true' in the sense of having predictability, making claims that most people would agree upon, like having a three-coloured coat (but didn't people once agree that shrews were dangerous? Wasn't that a claim about predictability at the time?)

But the authority of science also allows its claims to appear to be more true than it is. Among other things, it tends to name animals as passive, or instinct-driven, as not having autonomy or rationality, and so on, as I have argued.

Yet there are other stories to tell. Some of these are better stories, I would assert, because they do a better job of describing the kinds of animals and their behaviour that most of us experience in our everyday lives, the animals of 'common sense'. Even within science, the growing concern to think about animal cognition begins to grant them more autonomy and lives of their own, even if it is inevitably limited by the constraints of scientific methods. Outside of that narrow framework, there are many people whose experiences of animals in human culture simply do not fit into the straitjacket of scientific descriptions – pet owners, animal trainers, sheep farmers, people with disabilities who rely on assistance animals, and so on.

People who work with animals every day, such as dog breeders for example, tend to have a complex relationship to the discourses and knowledge of science. They may, for instance, make use of knowledge of genetics, taking some of that information on board while rejecting the rest. They may, similarly, incorporate tales of canine instincts and wolf ancestry into the ways they account for the behaviour of the dogs, while simultaneously talking about the idiosyncrasies and flexibility of 'their' dogs in relationship to themselves (Smart 1993).

The much-publicized concern about public (lack of) understanding of science, following the Royal Society's report in 1985, has tended to focus on the deficiencies of 'the public', their failures to understand what scientists are trying to tell them. The rhetoric of failure and deficiency comes into public pronouncements by scientists about the need for animals in research; the public must be better informed. Thus a letter to the *British Medical Journal*, defending animal use, remarked that

> It is now time for the public to be continually reminded by doctors that the tremendous developments of modern medicine in the past 50 years . . . would not be available but for animal experimental work.
>
> (Drury *et al.* 1990)

Similarly, Mark Matfield, director of Britain's Research Defence Society, suggested that the medical benefits

> are quite apparent to those engaged in biomedical research but are not appreciated by the general public at large. This lack of understanding leaves them prey to the propaganda currently being produced by the animal rights movement.
>
> (Matfield 1989)

From the perspective of those dedicated to defending the use of animals in medical research, there is no doubt a need to make the public 'better informed'. Perhaps they are 'prey to propaganda'. But this kind of rhetoric also reinforces an image of people outside science as somehow lacking, ignorant – an image which is not sustained by research into public perceptions of science. On the contrary, recent studies of how people see science and scientists suggest that they are often quite sophisticated in the ways they make use of – or reject – prevailing scientific orthodoxies. Sheep farmers on the Cumbrian fells have detailed knowledge of local ecosystems and sheep behaviour, details which were ignored by government experts pronouncing on radioactive fallout after Chernobyl; not surprisingly, the farmers were sceptical of expert opinion (Wynne 1991).

If we are to take seriously a view of non-human animals that respects them, then we must learn to understand them. One source of understanding is clearly the animals themselves. But another, often overlooked, source of knowledge is the people who know those animals best; people who work daily with animals often take scientific stories into account. They do so, however, by seeing how the scientific stories stand up against their own working knowledges of the animals. If the science fails on this test, then it will simply not be incorporated. That is not about ignorance of science; it is about an understanding of nature, of animals, that science has ignored.

By listening to these other accounts, human and non-human, I believe we can improve our scientific stories. A science that is responsive to, and prepared to incorporate, other ways of telling stories is, ironically, more likely to be a science that is 'objective' in the sense of telling a better story; this is similar to the way in which Sandra Harding distinguishes between a 'weak objectivity' that characterizes science as we know it, and the need for a strong objectivity that would include multiple accounts from different experiences (Harding 1992).

According other creatures more respect, and listening to other stories, means thinking beyond the narrow frameworks of reductionist science. It means, for instance, asking questions about the complex social experiences of a creature's life, rather than slotting it into a box and assuming that whatever it becomes is the product of genes or hormones. If scientific stories were improved, then the lives of animals might also be improved – particularly the lives of animals in science itself. How might the practices and assumptions of science be changed if the white rat were seen as special?

At a conference in Montreal in 1992, I listened to accounts of what wonderful things assistance animals had done for humans (for which special awards were being given). These are tales of dogs that rushed behind and lay under the wheelchair to stop it from slipping, of horses that adjust their movements to cater for the specific movements of children with severe physical disabilities. And how did the dog know (as human observers said she did) that the comatose child could be awakened, like the Sleeping Beauty, with a kiss? Scientific accounts would no doubt claim it was just coincidence, that the dog just tended to lick sleepers' faces. Most people, I suspect, would be as moved as I was by these stories, and would prefer to believe the testimony of those who actually saw and knew those animals. This is not about 'failure to understand' science. If people prefer another interpretation, then it is much more likely that they do so because they reject the narrowness and irrelevance of the scientific story – and because of their deep respect for animals' abilities.

Incorporating wider understandings, telling different stories, would also improve our own lives as human animals. Partly it would do so because it would give us a much richer sense of ourselves *and* of our relationships with non-humans if we saw them as complex, beautiful and interesting (by contrast to the impoverished view of ourselves that inevitably emerges from seeing human animals as just another bunch of instincts). Partly, too, it would make recourse to biologically determinist accounts redundant. The reason why I want to challenge those accounts is not only that they impoverish our views of ourselves, but that they also support views of animals as automata. The two images are but two sides of a coin.

Notes

1 Benefit here is relative. The welfare of the animal is, in many contexts, dependent on its value in the marketplace. Veterinary practice in relation to agriculture is a good example.

2 Non-human primates are an interesting case here. As Donna Haraway has noted (1991), recent work in primatology begins to transgress and challenge some of the rigid boundaries of previous scientific accounts. This is perhaps unsurprising; monkeys and apes are indeed at the boundaries of human culture and 'nature', but there is much less evidence of boundary transgressions for non-primate species.

3 I am not convinced by arguments such as Sharon Marcus (1992: 400) that rape is better understood as a 'form of invagination in which rape scripts the female body'. I accept that the discourses of rape are problematic, and can posit women as victims. But to talk of scripting bodies seems to me to move away from the real suffering that rape produces. By analogy, I insist that we must take animal suffering into account, and not see it as discourses scripting animal bodies.

REFERENCES

Adams, C. (1990) *The Sexual Politics of Meat*. Cambridge: Polity/Blackwell.

Alic, M. (1984) *Hypatia's Heritage*. London: The Women's Press.

Allen, L. and Gorski, R. (1992) Sexual orientation and the size of the anterior commissure in the human brain, *Proceedings of the National Academy of Sciences of the USA*, 89, 7199–202.

Arluke, A. (1988) Sacrificial symbolism in animal experimentation: Object or pet? *Anthrozoos*, 2, 97–116.

Arluke, A. (1990) Uneasiness among animal technicians, *Laboratory Animal*, 19, 20–39.

Arluke, A. (1992a) 'The ethical culture of primate labs', paper given at Science and the Human–Animal Relationship meeting, Amsterdam, March.

Arluke, A. (1992b) Trapped in a guilt cage, *New Scientist*, 134 (1815), 33–35.

Bailey, M.B. (1986) Every animal is the smartest: Intelligence and the ecological niche, in R.J. Hoage and L. Goldman (eds) *Animal Intelligence: Insights into the Animal Mind*. Washington DC: Smithsonian Institution Press.

Barclay, R., Herbert, W.J. and Poole, T.B. (1988) The disturbance index: A behavioural method of assessing the severity of common laboratory procedures on rodents, UFAW Animal Welfare Research Report, No. 2, Potters Bar: Universities Federation of Animal Welfare.

Barlow, G. (1989) Has sociobiology killed ethology, or revitalized it? in P.P.G. Bateson and P.H. Klopfer (eds) *Perspectives in Ethology: Vol 8: Whither Ethology?* London: Plenum Press.

Bateson, P.P.G. and Klopfer, P.H. (1992) Assessment of pain in animals, in M.S. Dawkins and M. Gosling (eds) *Ethics in Research on Animal Behaviour*. London: Academic Press/Association for the Study of Animal Behaviour.

Bayliss, L.E. (1966) *Living Control Systems*. London: English Universities Press.

Bazerman, C. (1988) *Shaping Written Knowledge: The Genre and Activity of the Experimental Article in Science*. Madison: University of Wisconsin Press.

Bekoff, M. and Jamieson, D. (1991) Reflective ethology, applied philosophy and the moral status of animals, *Perspectives in Ethology*, 9, 1–47.

Belenky, M.B., Clinchy, B.M., Goldberger, N.R. and Tarule, J.M. (1986) *Women's Ways of Knowing: The Development of Self, Voice and Mind*. New York: Basic Books.

Benney, N. (1983) All of one flesh, in L. Caldecott and S. Leland (eds) *Reclaim the Earth*. London: The Women's Press.

Benton, T. (1988) Humanism = speciesism: Marx on humans and animals, *Radical Philosophy*, 50, 4–18.

Benton, T. (1991) Biology and social science: Why the return of the repressed should be given a (cautious) welcome, *Sociology*, 25, 1–29.

Benton, T. (1993) *Unnatural Relations*. London: Verso.

Biology and Gender Study Group (1989) The importance of feminist critique for contemporary cell biology, in N. Tuana (ed.) *Feminism and Science*. Bloomington: Indiana University Press.

Birke, L. (1980) From zero to infinity: Scientific views of lesbians, in Brighton Women and Science Group (eds) *Alice Through the Microscope*. London: Virago.

Birke, L. (1986) *Women, Feminism and Biology: The Feminist Challenge*. Brighton: Wheatsheaf.

Birke, L. (1989) How do gender differences in behaviour develop? A reanalysis of the role of early experience, in P.P.G. Bateson and P.H. Klopfer (eds) *Perspectives in Ethology: Vol. 8 Whither Ethology?* London: Plenum.

Birke, L. (1991a) Science, feminism and animal natures: I: Extending the boundaries, *Women's Studies International Forum*, 14, 443–9.

Birke, L. (1991b) Science, feminism and animal natures: II: Feminist critiques and the place of animals in science, *Women's Studies International Forum*, 14, 451–8.

Birke, L. (1992) In pursuit of difference, in L. Keller and G. Kirkup (eds) *Inventing Women*. London: Polity Press.

Birke, L. and Michael, M. (1992a) The researchers dilemma, *New Scientist*, 4 April, 25–28.

Birke, L. and Michael, M. (1992b) Views from behind the barricade, *New Scientist*, 4 April, 29–32.

Birke, L. and Vines, G. (1987) Beyond nature versus nurture: Process and biology in the development of gender, *Women's Studies International Forum*, 10, 555–70.

Bleier, R. (1984) *Science and Gender*. Oxford: Pergamon.

Bordo, S. (1990) Reading the slender body, in M. Jacobus, E.F. Keller and S. Shuttleworth (eds) *Body/Politics: Women and the Discourses of Science*. London: Routledge.

British Association for the Advancement of Science (1993) *Animals and the Advancement of Science*. London: BAAS.

British Society for Social Responsibility in Science (BSSRS) Sociobiology Group (1984) in L. Birke and J. Silvertown (eds) *More than the Parts: Biology and Politics*. London: Pluto Press.

Brown, A. (1978) *Who Cares for Animals? 150 Years of the RSPCA*. London: Heinemann.

Brown, J. (1875) Letter, *The Spectator*, 26 June, 817.

Burt, M.R. (1988) The animal as alter ego: Cruelty, altruism and the work of art, in A. Rowan (ed.) *Animals and People Sharing the World*. London: University Press of New England.

Caldecott, L. and Leland, S. (eds) (1982) *Reclaim the Earth*. London: The Women's Press.

Candland, D.K. (1987) Tool Use, in G. Mitchell and J. Erwin (eds) *Comparative Primate Biology: Vol. 2, Part B: Behavior, Cognition and Motivation*. New York: Alan R. Liss.

Carruthers, P. (1986) *Introducing Persons: Theories and Arguments in the Philosophy of Mind*. London: Croom Helm.

Chamove, A., Anderson, J.R., Morgan-Jones, S.C. and Jones, S.P. (1982) Deep woodchip litter: Hygiene, feeding and behavioral enhancement in eight primate species, *International Journal for the Study of Animal Problems*, 31, 308–18.

Cheyney, D.L. and Seyfarth, R.M. (1990) *How Monkeys See the World: Inside the Mind of Another Species*. Chicago: University of Chicago Press.

Chodorow, N. (1979) *The Reproduction of Mothering*. Berkeley: University of California Press.

Chow, R. (1989) 'It's you, and not me': Domination and 'othering' in theorizing the 'Third World', in E. Weed (ed.) *Coming to Terms: Feminism, Theory, Politics*. London: Routledge.

Chrystos (1983) No rock scorns me as whore, in C. Moraga and G. Anzaldua (eds) *This Bridge Called My Back: Writings by Radical Women of Color*. New York: Kitchen Table Press.

Clark, S. (1982) *The Nature of the Beast*. Oxford: Oxford University Press.

Clutton-Brock, J. (1987) *A Natural History of Domesticated Animals*. Cambridge: Cambridge University Press.

Code, L. (1989) Experience, knowledge and responsibility, in A. Garry and M. Pearsall (eds) *Women, Knowledge and Reality: Explorations in Feminist Philosophy*. London: Unwin Hyman, 157–72.

Coleman, W. (1977) *Biology in the Nineteenth Century: Problems of Form, Function and Transformation*. Cambridge: Cambridge University Press.

Collard, A. (with J. Contrucci) (1988) *The Rape of the Wild*. London: The Women's Press.

Collins, P.H. (1991) *Black Feminist Thought*. London: Unwin Hyman.

Corbet, G. and Ovenden, D. (1980) *The Mammals of Britain and Europe*. London: Collins.

Cox, C. (1993) Ecofeminism, in L.S. Keller and G. Kirkup (eds) *Inventing Women*. London: Polity Press.

Crisler, L. (1991) Life with a wolf pack, in L. Anderson (ed.) *Sisters of the Earth*. New York: Vintage.

Crowther, B. (1994) Towards a feminist critique of television natural history programmes, in D. Reynolds and P. Florence (eds) *Feminist Subjects: Multimedia*. London: Longman.

Daly, M. (1979) *Gyn/Ecology: The Metaethics of Radical Feminism*. Boston: Beacon Press.

Davidson, A.I. (1991) The Horror of Monsters, in J. Sheehan and M. Sosna (eds) *The Boundaries of Humanity: Humans, Animals, Machines*. Berkeley: University of California Press.

Davis, H. and Memmott, J. (1982) Counting behavior in animals: A critical evaluation, *Psychology Bulletin*, 92, 547–71.

Dawkins, M.S. (1980) *Animal Suffering: The Science of Animal Welfare*. London: Chapman and Hall.

De Moubray, J. (1987) *The Thoroughbred Business*. London: Hamish Hamilton.

De Pomerai, D. (1985) *From Gene to Animal*. Cambridge: Cambridge University Press.

Diamond, C. (1992) The importance of being human, in D. Cockburn (ed.) *Human Beings*. Cambridge: Cambridge University Press.

Diamond, C. and Orenstein, G.F. (1990) *Reweaving the World: the Emergence of Ecofeminism*. Sierra Club Books, San Francisco.

Donovan, J. (1990) Animal rights and feminist theory, *Signs*, 15, 350–75.

Doubiago, S. (1989) Mama Coyote talks to the boys, in J. Plant (ed.) *Healing the Wounds: The Promise of Ecofeminism*. London: Green Print.

Doyal, L. and Gough, I. (1991) *A Theory of Human Need*. London: Pluto.

Driscoll, J. and Bateson, P. (1992) Animals in Behavioural Research, in M.S. Dawkins and M. Gosling (eds) *Ethics in Research on Animal Behaviour*. London: Academic Press/Association for the Study of Animal Behaviour.

Drury, M., Wade, O., Kuenssberg, E., *et al.* (1990) Benefits of medical research and the doctor's responsibility, *British Medical Journal*, 300, 538.

Elia, I. (1985) *The Female Animal*. Oxford: Oxford University Press.

Elliott, P. (1987) Vivisection and the emergence of experimental physiology in nineteenth century France, in N. Rupke (ed.) *Vivisection in Historical Perspective*. London: Routledge.

Elston, M.A. (1987) Women and anti-vivisection in Victorian England, 1870–1900, in N. Rupke (ed.) *Vivisection in Historical Perspective*. London: Routledge.

Evans, E.P. (1987) *The Criminal Prosecution of Animals and Capital Punishment of Animals: The Lost History of Europe's Animal Trials*. London: Faber and Faber.

Farley, J. (1982) *Gametes and Spores: Ideas about Sexual Reproduction 1750–1914*. Baltimore: Johns Hopkins University Press.

Fausto-Sterling, A. (1985) *Myths of Gender*. New York: Basic Books.

Fausto-Sterling, A. (1989) Life in the XY corral, *Women's Studies International Forum*, 12, 319–31.

Fausto-Sterling, A. (1992) *Myths of Gender – Second edition*. New York: Basic Books.

Fee, E. (1983) Women's nature and scientific objectivity, in M. Lowe and R. Hubbard (eds) *Woman's Nature*. New York: Pergamon.

Fiddes, N. (1992) *Meat: A Natural Symbol*. London: Routledge.

Flax, J. (1990) Postmodernism and gender relations in feminist theory, in L. Nicholson (ed.) *Feminism/Postmodernism*. London: Routledge.

Fox, M.W. (1986) *Laboratory Animal Husbandry*. New York: State University of New York.

Fuss, D. (1989) *Essentially Speaking: Feminism, Nature and Difference*. London: Routledge.

Gatens, M. (1986) Feminism, philosophy and riddles without answers, in C. Pateman and E. Gross (eds) *Feminist Challenges: Social and Political Theory*. Australia: Allen and Unwin.

Gatens, M. (1991) A critique of the sex/gender distinction, in S. Gunew (ed.) *A Reader in Feminist Knowledge*. London: Routledge.

Gatens, M. (1992) Power, bodies and difference, in M. Barrett and A. Phillips (eds) *Destabilizing Theory: Contemporary Feminist Debates*. London: Polity Press.

Genova, J. (1989) Women and the mismeasure of thought, in N. Tuana (ed.) *Feminism and Science*. Bloomington: Indiana University Press.

Gilligan, C. (1982) *In a Different Voice: Psychological Theory and Women's Development*. Cambridge, MA: Harvard University Press.

Goodwin, B. (1984) Changing from an evolutionary to a generative paradigm in biology, in J.W. Pollard (ed.) *Evolutionary Theory: Paths into the Future*. Chichester: Wiley.

Gould, S.J. (1991) *Wonderful Life: The Burgess Shale and the Nature of History*. Harmondsworth: Penguin.

Gregory, B. (1990) *Inventing Reality: Physics as Language*. Chichester: Wiley.

Griffin, D.R. (1981) *The Question of Animal Awareness*. Los Altos: William Kaufman.

Griffin, D.R. (1992) *Animal Minds*. Chicago: Chicago University Press.

Gross, A.G. (1990) *The Rhetoric of Science*. Cambridge, MA: Harvard University Press.

Gross, M. and Averill, M.B. (1983) Evolution and patriarchal myths of scarcity and competition, in S. Harding and M. Hintikka (eds) *Discovering Reality: Feminist Perspectives on Epistemology, Metaphysics, Methodology and Philosophy in Science*. London: Reidel.

Guerrini, A. (1989) The ethics of animal experimentation in seventeenth-century England, *Journal of the History of Ideas*, 50, 391–408.

Halpin, Z.T. (1989) Scientific objectivity and the concept of 'the other', *Women's Studies International Forum*, 12, 285–94.

Hamer, D.H., Hu, S., Magnuson, V.L., Hu, N. and Pattatucci, A.M.L. (1993) A linkage between DNA markers on the X chromosome and male sexual orientation, *Science*, 261, 321–7.

Haraway, D. (1989a) Situated knowledges: the science question in feminism and the privilege of partial perspective, *Feminist Studies*, 14, 575–99.

Haraway, D. (1989b) *Primate Visions*. London: Routledge.

Haraway, D. (1990) A manifesto for Cyborgs: Science, technology and socialist feminism in the 1980s, in L. Nicholson (ed.) *Feminism/Postmodernism*. London: Routledge, 190–233.

Haraway, D. (1991a) The contest for primate nature: daughters of man-the-hunter in the field, 1960–80, in D. Haraway *Simians, Cyborgs, and Women*. London: Free Association Books.

Haraway, D. (1991b) The biopolitics of postmodern bodies, in D. Haraway, *Simians, Cyborgs and Women*. London: Free Association Books.

Harding, S. (1986) *The Science Question in Feminism*. Milton Keynes: Open University Press.

Harding, S. (1991) *Whose Science? Whose Knowledge?* Buckingham: Open University Press.

Harding, S. (1992) Feminist justificatory strategies, in A. Garry and M. Pearsall (eds) *Women, Knowledge and Reality: Explorations in Feminist Philosophy*. London: Routledge.

Harper, K. (1986) *Give Me My Father's Body*. Liqualuit: Blacklead Books.

Harrison, P. (1989) Theodicy and animal pain, *Philosophy*, 64, 79–92.

Hartsock, N. (1983) The Feminist Standpoint: Developing the ground for a specifically feminist historical materialism, in S. Harding and M. Hintikka (eds) *Discovering Reality: Feminist Perspectives on Epistemology, Metaphysics, Methodology and Philosophy in Science*. London: Reidel.

Hawkesworth, M. (1989) Knowers, knowing, known: Feminist theory and claims of truth, *Signs: Journal of Women in Culture and Society*, 14, 533–57.

Hearne, V. (1987) *Adam's Task: Calling Animals by Name*. London: Heinemann.

Hekman, S.J. (1990) *Gender and Knowledge: Elements of a Postmodern Feminism*. Boston: Northeastern University Press.

Horigan, S. (1990) *Nature and Culture in Western Discourses*. London: Routledge.

Houston, B. (1987) Rescuing womanly virtues: Some dangers of moral reclamation, in M. Hanen and K. Nielsen (eds) *Science, Morality and Feminist Theory*. Calgary: University of Calgary Press.

Hubbard, R. (1990) *The Politics of Women's Biology*. New Brunswick and London: Rutgers University Press.

Hubbard, R. and Wald, E. (1993) *Exploding the Gene Myth*. Boston: Beacon Press.

Hull, D. (1988) *Science as a Process: An Evolutionary Account of the Social and Conceptual Development of Science*. Chicago: University of Chicago Press.

Hull, D. (1992) The effect of essentialism on taxonomy: Two thousand years of stasis, in M. Ereshevsky (ed.) *The Units of Evolution: Essays on the Nature of Species*. Cambridge, MA: MIT Press.

Jasper, J. and Nelkin, D. (1992) *The Animal Rights Crusade*. New York: The Free Press.

Johnson, W. (1850) The morbid emotions of women, quoted in P. Jalland and J. Hooper (eds) *Women from Birth to Death* (1986). Brighton: Harvester.

Jones, G. (1980) *Social Darwinism and English Thought: The Interaction between Biological and Social Theory*. Brighton: Harvester.

Jordanova, L. (1991) *Sexual Visions*. Brighton: Harvester.

Keller, E.F. (1980) Baconian Science: A Hermaphroditic Birth, *The Philosophical Forum*, 11, 299–308.

Keller, E.F. (1983) *A Feeling for the Organism*. San Francisco: W.H. Freeman & Co.

Keller, E.F. (1985) *Perspectives on Gender and Science*. New Haven, CT: Yale University Press.

Keller, E.F. (1992) Feminism and Science, in A. Garry and M. Pearsall (eds) *Women, Knowledge and Reality*. London: Routledge.

Keller, E.F. (1993) *Secrets of Life, Secrets of Death: Essays on Language, Gender and Science*. London: Routledge.

Kennedy, J.S. (1992) *The New Anthropomorphism*. Cambridge: Cambridge University Press.

Kite, S. (1990) *Secret Suffering: Inside a British Laboratory*. London: British Union for the Abolition of Vivisection.

Klinkenborg, G. (1993) Barnyard diversity. *Audubon*, 95, 78–88.

Knorr-Cetina, K. (1983) The ethnographic study of scientific work: towards a constructionist interpretation of science, in K. Knorr-Cetina and M. Mulkay (eds) *Science Observed: Perspectives on the Social Study of Science*. London: Sage.

Kuhn, A. (1988) The body and cinema: Some problems for feminism, in S. Sheridan (ed.) *Grafts: Essays in Feminist Cultural Theory*. London: Verso.

Lane-Petter, W. (1953) Uniformity in laboratory animals, *Laboratory Practice*, 1, 30–33.

Lansbury, C. (1985) *The Old Brown Dog: Women, Workers and Vivisection in Edwardian England*. Madison: University of Wisconsin Press.

Laqueur, T. (1990) *Making Sex: Body and Gender from the Greeks to Freud*. Cambridge, MA: Harvard University Press.

Latour, B. (1983) Give me a laboratory and I will raise the world, in K. Knorr-Cetina and M. Mulkay (eds) *Science Observed*. London: Sage.

Latour, B. (1987) *Science in Action: How to Follow Engineers in Society*. Milton Keynes: Open University Press.

Latour, B. and Woolgar, S. (1979) *Laboratory Life: The Construction of Scientific Facts*. London: Sage.

Lawrence, E. (1985) *Hoofbeats: Studies of Human–Horse Interactions*. Bloomington: University of Indiana Press.

Leahy, M.P.T. (1991) *Against Liberation: Putting Animals in Perspective*. London: Routledge.

Lederer, S. (1992) Political animals: The shaping of biomedical research literature in twentieth century America, *Isis*, 83, 61–79.

LeGuin, U. (1987a) She Unnames Them, in U. LeGuin, *Buffalo Gals and Other Presences*. Santa Barbara: Capra Press.

LeGuin, U. (1987b) Mazes, in U. LeGuin, *Buffalo Gals and Other Presences*. Santa Barbara: Capra Press.

LeGuin, U. (1987c) The Author of the Acacia Seeds, in U. LeGuin, *Buffalo Gals and Other Presences*. Santa Barbara: Capra Press.

LeGuin, U. (1989) Women/wilderness, in J. Plant (ed.), *Healing the Wounds: the Promise of Ecofeminism*. London: Green Print.

Lennox, A. (1987) *Greyhounds: The Sporting Breed*. London: Sportsmans Press.

LeVay, S. (1993) *The Sexual Brain*. Cambridge, MA: MIT Press.

Lloyd, G. (1984) *The Man of Reason: 'Male' and 'Female' in Western Philosophy*. London: Methuen.

Longino, H.E. (1989) Can there be a feminist science?, in N. Tuana (ed.) *Feminism and Science*. Bloomington: Indiana University Press.

Lovejoy, E.O. (1936) *The Great Chain of Being*. New York: Harper & Row.

Lynch, M. (1985) *Art and Artifact in Laboratory Science: A Study of Shop Work and Shop Talk in a Research Laboratory*. London: Routledge.

Lynch, M.E. (1988) Sacrifice and the transformation of the animal body into a scientific object: Laboratory culture and ritual practice in the neurosciences, *Social Studies of Science*, 18, 265–89.

McClintock, M. and Adler, N. (1979) The role of the female during copulation in wild and domestic rats (*Rattus norvegicus*), *Behaviour*, 67, 67–96.

McConway, K. (1992) The number of animals in animal behaviour experiments: Is Still still right? in M.S. Dawkins and M. Gosling (eds) *Ethics in Research on Animal Behaviour*. London: Academic Press/Association for the Study of Animal Behaviour.

McFarland, D. (1989) *Problems of Animal Behaviour*. Harlow: Longman Scientific.

McGaughey, C.A., Thompson, H.V. and Chitty, D. (1947) The Norway Rat, in A. Worden (ed.) *UFAW Handbook on the Care and Management of Laboratory Animals*. London: Universities Federation for Animal Welfare/Bailliere.

McGrew, W. (1992) *Chimpanzee Material Culture: Implications for Human Evolution*. Cambridge: Cambridge University Press.

Mackenzie, B.D. (1977) *Behaviourism and the Limits of Scientific Method*. London: Routledge.

Maehle, A.H. and Trohler, U. (1987) Animal experimentation from antiquity to the end of the eighteenth century: attitudes and arguments, in N. Rupke (ed.) *Vivisection in Historical Perspective*. London: Routledge.

Mansfield, A. and McGinn, B. (1993) Pumping Irony: The Muscular and the Feminine, in S. Scott and D. Morgan (eds) *Body Matters*. London: Falmer Press.

Marcus, S. (1992) Fighting bodies, fighting words: a theory and politics of rape prevention, in J. Butler and J. Scott (eds) *Feminists Theorize the Political*. London: Routledge.

Martin, E. (1989) *The Woman in the Body*. Milton Keynes: Open University Press.

Martin, E. (1995) Working across the Human-Other Divide, in R. Hubbard and L. Birke (eds) *Reinventing Biology*. Bloomington: University of Indiana Press.

Mason, P. (1990) *Deconstructing America: Representations of the Other*. London: Routledge.

Masters, J.C. (1995) rEvolutionary Theory: Reinventing Our Origin Myths, in R. Hubbard and L. Birke (eds) *Reinventing Biology*. Bloomington, University of Indiana Press.

Matfield, M. (1989) The animal rights threat to biomedical research, *Biologist* 36, 238–40.

Merchant, C. (1982) *The Death of Nature: Women, Ecology and the Scientific Revolution*. London: Wildwood.

Midgley, M. (1978) *Beast and Man*. Brighton: Harvester.

Mills, P.J. (1991) Feminism and Ecology: On the domination of nature, *Hypatia*, 6, 162–78.

Moir, A. and Jessel, D. (1989) *BrainSex: The Real Differences between Men and Women*. London: Michael Joseph.

Moore, C.L. (1984) Development of mammalian sexual behavior, in E.S. Gollin (ed.) *The Comparative Development of Adaptive Skills: Evolutionary Implications*. Hillsdale, NJ: Erlbaum.

Morgan, D. (1993) You too can have a body like mine: Reflections on the male body and masculinities, in S. Scott and D. Morgan (eds) *Body Matters*. London: Falmer Press.

Morgan, L., quoted in G. Jones (1980) *Social Darwinism and English Thought*. Brighton: Harvester.

Morton, D. and Griffiths, P. (1985) Guidelines on the recognition of pain, distress and discomfort in experimental animals and an hypothesis for assessment, *Veterinary Record*, 116, 431–36.

Nelkin, D. and Jasper, J. (1992) The animal rights controversy, in D. Nelkin (ed.) *Controversy: Politics of Technical Decisions*. London: Sage.

Norwood, V. (1993) *Made from this Earth: American Women and Nature*. Chapel Hill: University of North Carolina Press.

Noske, B. (1989) *Humans and Other Animals*. London: Pluto Press.

Oyama, S. (1985) *The Ontogeny of Information*. Cambridge: Cambridge University Press.

Passmore, J. (1974) *Man's Responsibility for Nature*. London: Duckworth.

Perec, G. (1991) *Cantatrix sopranica L. et autres écrits scientifiques*. Paris: Editions du Seuil.

Petchesky, R. (1987) Foetal images: The power of visual culture in the politics of reproduction, in M. Stanworth (ed.) *Reproductive Technologies*. London: Polity Press.

Plumwood, V. (1991) Nature, Self and Gender: Feminism, environmental philosophy and the critique of rationalism, *Hypatia*, 6, 3–27.

Potter, E. (1988) Modeling the gender politics in science, in N. Tuana (ed.) *Feminism and Science*. Bloomington: Indiana University Press.

Premack, D. and Premack, A.J. (1983) *The Mind of an Ape*. New York: Norton.

Quinn, M.S. (1993) Corpulent cattle and milk machines, *Society and Animals*, 1, 145–58.

Rachels, J. (1990) *Created from Animals: The Moral Implications of Darwinism*. Oxford: Oxford University Press.

Radner, D. and Radner, M. (1989) *Animal Consciousness*. New York: Prometheus Books.

Regan, T. (1983) *The Case for Animal Rights*. London: Routledge.

Research Defence Society (1919) Quarterly Report, October. Wellcome Library.

Richardson, R. (1989) *Death, Dissection and the Destitute*. London: Pelican.

Rifkin, J. (1992) *Beyond Beef: The Rise and Fall of the Cattle Culture*. New York: Dutton.

Riley, D. (1988) *Am I That Name? Feminism and the Category of 'Women' in History*. London: Macmillan.

Ritvo, H. (1987) *The Animal Estate: The English and other Creatures in the Victorian Age*. Harmondsworth: Penguin.

Ritvo, H. (1990) The power of the word: Scientific nomenclature and the spread of Empire. *Victorian Newsletter*, Spring, 5–8.

Ritvo, H. (1991) The animal connection, in J. Sheehan and M. Sosna (eds) *The Boundaries of Humanity: Humans, Animals, Machines*. Berkeley: University of California Press.

Rodd, R. (1990) *Biology, Ethics and Animals*. Oxford: Clarendon Press.

Rogers, L. (1995) They are *only* animals, in R. Hubbard and L. Birke (eds) *Reinventing Biology*. Bloomington: Indiana University Press.

Rollin, B. (1989) *The Unheeded Cry: Animal Consciousness, Animal Pain and Science*. Oxford: Oxford University Press.

Rollin, B. (1992) *Animal Rights and Human Morality*. Buffalo: Prometheus Books.

Romer, A. (1970) *The Vertebrate Body*. Philadelphia: Saunders.

Rose, H. (1983) Hand, brain and heart: A feminist epistemology for the natural sciences, *Signs*, 9, 73–90.

Rose, H. (1994) *Love, Power and Knowledge: Towards a Feminist Transformation of the Sciences*. Cambridge: Polity.

Rose, M. (1988) Henderson, in I. Zahava (ed.) *Through Other Eyes: Animal Stories by Women*. Freedom, California: Crossing Press.

Rose, S. (1991) Proud to be speciesist, *New Statesman and Society*, 26 April, 21.

Rosenberg, R. (1982) *Beyond Separate Spheres: Intellectual Roots of Modern Feminism*. New Haven, CT: Yale University Press.

Rosser, S. (1989) Feminist scholarship in the sciences: Where are we now and when can we expect a theoretical breakthrough?, in N. Tuana (ed.) *Feminism and Science*. Bloomington: Indiana University Press.

Rothschild, M. (1986) *Animals and Man: The Romanes Lecture 1984–5*. Oxford: Clarendon Press.

Rowan, A. (ed.) (1988) *Animals and People Sharing the World*. Hanover: University Press of New England.

Ruddick, S. (1982) Maternal thinking, *Feminist Studies*, 6, 342–69.

Ruether, R.R. (1989) Toward an ecological-feminist theory of nature, in J. Plant (ed.) *Healing the Wounds: The Promise of Ecofeminism*. Green Print: London.

Rupke, N. (1987) Pro-vivisection in England in the early 1880s: Arguments and motives, in N. Rupke (ed.) *Vivisection in Historical Perspective*. London: Routledge.

Russell, N. (1986) *Like Engend'ring Like: Heredity and Animal Breeding in Early Modern England*. Cambridge: Cambridge University Press.

Russett, C.R. (1989) *Sexual Science*. Cambridge, MA: Harvard University Press.

Ryden, H. (1972) *Mustangs: A Return to the Wild*. New York: Viking.

Ryder, R. (1989) *Animal Revolution: Changing Attitudes towards Speciesism*. Oxford: Blackwell.

Sales, G. (1988) Effects of environmental ultrasound on behaviour of laboratory rats, in *Laboratory Animal Welfare Research: Rodents*. Potters Bar: Universities Federation for Animal Welfare.

Saxton, M. (1984) Born and unborn: The implications of reproductive technologies for people with disabilities, in R. Arditti, R. Duelli Klein and S. Minden (eds) *Test-tube Women: What Future for Motherhood?* London: Pandora.

Schiebinger, L. (1989) *The Mind has no Sex? Women in the Origins of Modern Science*. Cambridge, MA: Harvard University Press.

Schupback, W. (1987) A select iconography of animal experiment, in N. Rupke (ed.) *Vivisection in Historical Perspective*. London: Routledge.

Scruton, R. (1986) *Sexual Desire*. London: Weidenfeld and Nicholson.

Serpell, S. (1986) *In the Company of Animals*. Oxford: Blackwell.

Serpell, J. (1988) Pet-keeping in non-Western societies: Some popular misconceptions, in A. Rowan (ed.) *Animals and People Sharing the World*. Hanover: UNE.

Shaw, E. and Darling, J. (1985) *Female Strategies*. New York: Walker.

Shilling, C. (1993) *The Body and Social Theory*. London: Sage.

Shiva, V. (1989) *Staying Alive: Women, Ecology and Development*. London: Zed Press.

Singer, P. (1975) *Animal Liberation*. London: Jonathan Cape.

Slicer, D. (1991) Your daughter or your dog?, *Hypatia*, 6, 108–24.

Smart, K. (1993) *Resourcing ambivalence: dog breeders, animals and the social studies of science*. Ph.D. dissertation, School of Independent Studies, University of Lancaster.

Smith, J., Birke, L. and Sadler, D. (1994) Reporting animal use in scientific papers, *Laboratory Animals* (in press).

Snow, D., Persson, S. and Rose, R. (1983) *Equine Exercise Physiology*. Cambridge: Granta.

Spelman, E. (1982) Woman as body: Ancient and contemporary views, *Feminist Studies*, 8, 109–31.

Spretnak, C. (1989) Toward an ecofeminist spirituality, in J. Plant (ed.) *Healing the Wounds: The Promise of Ecofeminism*. London: Green Print.

Spretnak, C. (1990) Ecofeminism: Our Roots and Flowering, in I. Diamond and G. Orenstein (eds) *Reweaving the World: The Emergence of Ecofeminism*. San Francisco: Sierra Club Books.

Starhawk (1990) Power, Authority and Mystery: Ecofeminism and Earth-based Spirituality, in I. Diamond and G. Orenstein (eds) *Reweaving the World: The Emergence of Ecofeminism*. San Francisco: Sierra Club Books.

Still, A. (1982) On the number of subjects used in animal behaviour experiments, *Animal Behaviour*, 30, 873–80.

Stinson, S. (1990) Whale and Woman, in T. Corrigan and S. Hoppe (eds) *And a Deer's Ear, Eagle's Song and Bear's Grace: Animals and Women*. San Francisco, Cleis Press.

Sydie, R.A. (1987) *Natural Women, Cultured Men: A Feminist Perspective on Sociological Theory*. Milton Keynes: Open University Press.

Tester, K. (1991) *Animals and Society: The Humanity of Animal Rights*. London: Routledge.

Thiele, B. (1986) Vanishing acts in social and political thought: Tricks of the trade, in C. Pateman and E. Gross (eds) *Feminist Challenges: Social and Political Theory*. Sydney: Allen and Unwin.

Thomas, K. (1983) *Man and the Natural World*. Harmondsworth: Penguin.

Thomas, R.K. (1986) Vertebrate intelligence: A review of the laboratory research, in R.J. Hoage and L. Goldman (eds) *Animal Intelligence: Insights into the Animal Mind*. Washington DC: Smithsonian Institution.

Tuana, N. (1989) The weaker seed: The sexist bias of reproductive theory, in N. Tuana (ed.) *Feminism and Science*. Bloomington: Indiana University Press.

Walker, A. (1988) Am I Blue?, in I. Zahava (ed.) *Through Other Eyes: Animal Stories by Women*. Freedom, California: Crossing Press.

Walker, S. (1983) *Animal Thought*. London: Routledge and Kegan Paul.

Warren, K.J. and Cheney, J. (1991) Ecological feminism and ecosystem ecology, *Hypatia*, 6, 179–97.

Wieder, D.L. (1980) Behavioristic operationalism and the life-world: Chimpanzees and the chimpanzee researchers in face-to-face interaction, *Sociological Inquiry*, 50, 75–103.

Willett, P. (1966) *The Thoroughbred*. London: Weidenfeld and Nicolson.

Williamson, T. and Bellamy, L. (1987) *Property and Landscape: A Social History of Land Ownership and the English Countryside*. London: George Philip.

Wittig, M. (1993) *The Straight Mind*. London: Harvester.

Worden, A.N. (1947) *UFAW Handbook on the Care and Management of Laboratory Animals*. London: Universities Federation for Animal Welfare/Bailliere.

Wright, S. (1922) The effects of inbreeding and crossbreeding on guineapigs, *US Department of Agriculture Bulletin*, No. 1090.

Wynne, B.E. (1991) Knowledges in context, *Science, Technology and Human Values*, 16, 111–21.

Wynne, B.E. (1992) Public understanding of science research: New horizons or hall of mirrors, *Public Understanding of Science*, 1, 37–43.

INDEX OF NON-HUMAN ANIMALS (BY COMMON NAMES)

INDEX